# The Illustrated Guide to Aerodynamics

## 2nd Edition

# The Illustrated Guide to Aerodynamics

## 2nd Edition

H.C. "Skip" Smith

**TAB Books**
**Division of McGraw-Hill, Inc.**
New York   San Francisco   Washington, D.C.   Auckland   Bogotá
Caracas   Lisbon   London   Madrid   Mexico City   Milan
Montreal   New Delhi   San Juan   Singapore
Sydney   Tokyo   Toronto

© 1992 by **TAB Books**.
TAB Books is a division of McGraw-Hill, Inc.

pbk          19 20  FGR/FGR   0 5 4
  hc   1 2 3 4 5 6 7 8 9 10  FGR/FGR  9 9 8 7 6 5 4 3 2

**Library of Congress Cataloging-in-Publication Data**

Smith, Hubert.
    The illustrated guide to aerodynamics / by Hubert "Skip" Smith.—
2nd ed.
      p.     cm.
    Includes bibliographical references (p. 327) and index.
    ISBN 0-8306-3901-2 (pbk.)      ISBN 0-8306-3902-0 (hard)
    1. Aerodynamics.  I. Title.
TL570.S56   1992
629.132'3—dc20                                         91-30240
                                                          CIP


Acquisitions Editor: Jeff Worsinger
Technical Editor: Norval G. Kennedy
Director of Production: Katherine G. Brown                    AV1
Book Design: Jaclyn J. Boone                                  3786

To my father, Col. Chas. S.,
the greatest guy I have ever known,
and the inspiration
for all of my accomplishments.

# Contents

# Acknowledgments

THIS BOOK WAS INSPIRED BY MY MANY STUDENTS in both engineering and aviation who sought a true understanding of the principles of flight. I would like to acknowledge their influence during my some 20 years of teaching at Penn State University and Embry-Riddle Aeronautical University.

I also thank David Thurston, noted aircraft designer and author, for reviewing the first few chapters and for his enthusiastic support of the work.

I would like to extend appreciation to the National Air and Space Museum of the Smithsonian Institution for use of their library and photographs, NASA Langley for their fine photographs, Domenic Maglieri of NASA Langley for his help in locating the photographs, and the staff of the Learning Resources Center at Embry-Riddle Aeronautical University for their courteous assistance.

I must also acknowledge the following organizations who supplied photographs: Aerolab, Beech Aircraft Corp., Cessna Aircraft Co., Gates-Learjet Corp., Lear Fan Limited, Mooney Aircraft Corp., Piper Aircraft Corp., and Raven Industries, Inc.

I am indebted to Karen Rider, my faithful typist, for making the task of writing easier.

I also owe a great deal of gratitude to my wife, Esther, for her patience, her criticism, her services as a model, and for accompanying me on numerous escapades to collect information and take photographs.

The second edition owes additional photo credits to: AMR-Avanti Sales N.A.; Analytical Methods Inc.; Boeing; General Dynamics-Convair Div.; LoPresti Piper Aircraft Engineering; and Northwest Airlines. I thank my capable secretary, Amy Myers, for her help in typing the second edition, and Denny Straussfogel for piloting duties while I made photos for the flight test examples. I also appreciate the efforts of my colleague, Dr. Mark Maughmer, for his technical assistance and help in locating references.

# Introduction

MOST FLIGHT STUDENTS WINCE at the mention of the term *aerodynamics*. Just plain aviation buffs avoid it like some ancient plague. Even engineering students often approach this mysterious subject with an attitude like that of an airsick-prone passenger boarding a small airplane. But aerodynamics is not a black art revealed only to a chosen few by some great god of flight. It can be understood by almost anyone, if properly explained. It can become interesting or even downright fascinating.

Aerodynamics is the science that deals with the motion of air and the forces on bodies moving through the air, according to official definition. For those of us interested in aeronautics, the term aerodynamics implies the explanation of why and how airplanes fly. It is also the key to the design of aircraft for efficient performance.

Because aerodynamics is a complex science, most books dealing with the subject are quite deep in scientific theory and use complicated mathematical relationships to explain it. Such complexity is necessary for those preparing to become professionally involved in the subject, such as scientists and engineers. On the other hand, pilots, technicians, and others with an interest in aviation have a need to understand the behavior of air and its effect on the airplane without getting into the exact numbers. Few books seem to treat aerodynamics in a sufficient manner for this group. As a result, many hangar stories and "old pilots' tales" abound about how and why airplanes fly as they do. Many of these tales are quite erroneous and mislead those involved in the safe and efficient handling of airplanes.

It is the purpose of this book to dispel some of these myths and to explain in everyday language just what happens to enable an airplane to fly, to perform as it does, and to be stable and controllable. It also shows, in the same straightforward manner, how these principles can be applied to the design of efficient airplanes.

This book could well serve as an introductory text for those entering the study of aeronautical engineering. As such, it could make the transition from layman to serious engineering student a much easier task. It should also be enlightening to almost anyone seriously interested in understanding the principles of flight.

I have taken a historical approach in explaining many of the subjects to put the technology in proper perspective. This approach also serves to make a bit more interesting reading. I have tried to explain principles as simply as possible, but have not tried to avoid any areas of importance simply because they are, by nature, somewhat complex. Occasionally you might encounter a few mathematical equations, but I assure you that they are very basic and should be understandable to even a high school student. They are used only when essential to the understanding of some important relations.

Finally, I must conclude this section with a word of caution. I want to emphasize that this book is intended to *explain* the principles of aerodynamics and design. It is *not* intended as a "do-it-yourself" manual for someone with no other appropriate background to design an airplane that will eventually be built and flown. Such endeavor takes many years of further study and experience. Neither the publisher nor I shall accept any responsibility for such use of the material in this book.

# 1

# What is aerodynamics?

THE HUMAN BEING IS SET APART from all other animals by an intelligent, reasoning mind. Another quality that humans seem to possess is an inherent drive to utilize that mind to achieve, to recognize challenges, and to attempt to conquer them. From the dawn of history, man has been lured to explore his world and to expand the horizons of his habitation. To accomplish this feat, chariots were devised, horses were harnessed, and sailing ships were created. In all of these endeavors, the idea of possibility surely must have been sparked by what was observed in nature. If deer and wild horses could travel great distances at high speeds, then man was tempted to try it. If fish could swim across oceans, then the possibility loomed for man to accomplish this feat.

## HISTORY OF FLIGHT

It would seem obvious, then, that the sight of birds flying through the air would entice men to attempt to do the same (Fig. 1-1). It is also not too surprising that the first attempts at human flight employed the concept of flapping wings. After all, this is the way that the birds did it. In early Greek mythology, we have accounts of men flying with birdlike wings. The story of Daedalus and his son Icarus utilized flight as a means of escape from the island of Crete where they were imprisoned. They were said to have created wings of feathers held together with wax. Icarus, however, being young and impetuous, flew too close to the sun, and the heat melted the wax in his wings as shown in Fig. 1-2. He plunged to the sea and suffered one of the first fatal crashes in history—at least according to mythology. Inaccurate as the scientific facts might be, this story points out the dream of flight that pervaded the human mind far back in history.

The concept of flapping wings occupied the thoughts of practically all those who dreamed of flight for many centuries. Notable among such dreamers was Leonardo da Vinci, best known for his artistic accomplishments. He was, however, a man of many talents and made significant contributions to the scientific knowledge of his day. He was obsessed with the idea of achieving flight by somehow transforming the muscle-power of man into lift and thrust through a flap-

Fig. 1-1. Primitive man contemplates flight by observing birds.

Fig. 1-2. Icarus falls from the sky as the sun melts the wax from his wings.

ping wing device such as that shown in Fig. 1-3. As talented as he was, he did not recognize the futility of this approach and he contributed little to the actual achievement of flight. Such efforts, however, did serve to perpetuate—and to encourage—the idea of human flight.

Fig. 1-3. An early ornithopter conceived by Leonardo da Vinci.

The first successful aircraft of any kind relied on an entirely different principle of physical science. Early experimenters observed smoke rising and even bits of ash and other material, which normally fell to the ground, were seen to rise in heated air. Joseph and Etienne Montgolfier took advantage of this phenomenon and designed and constructed the first hot air balloon. This took place in France in June of 1783 and is illustrated in Fig. 1-4. The Montgolfier brothers did not actually realize what caused their balloon to ascend. They thought at first that burning wood released some unknown gas that mysteriously caused objects to rise. The principle of lighter-than-air flight soon became known, however, and considerable activity in both hot-air and hydrogen-filled balloons took place before the end of 1783. It is somewhat amazing that modern balloons, such as shown in Fig. 1-5, work almost exactly as those devised by the Montgolfiers.

# AEROSTATICS

A balloon works on the *buoyancy principle*, the discovery of which is attributed to Archimedes. The pressure in any fluid, liquid or gas, increases with the depth. This is apparent if you dive into the sea and also if you descend from a high altitude in the air. In both cases the pressure gets greater as you go down. This is true of even small changes in height, although the change in pressure is, of course, very small also.

Figure 1-6 shows a small chunk of fluid within a larger container of the fluid. The pressure on the bottom surface is greater by an amount $\Delta P$ ($\Delta$ is pronounced *delta*, an abbreviation for difference) over the amount P on the top surface. If the

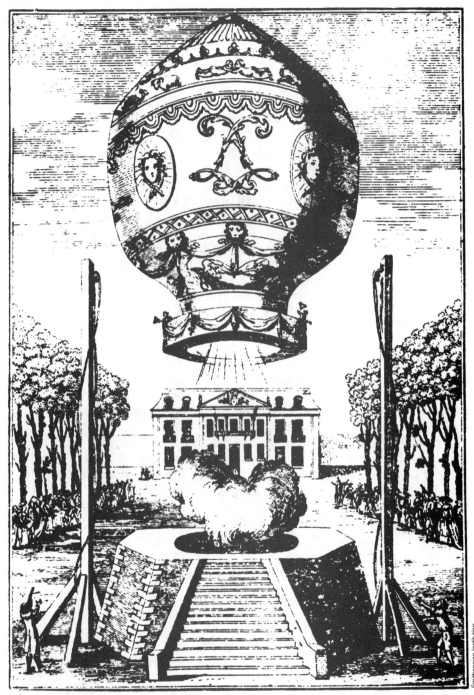

Fig. 1-4. Launching of the Montgolfier brothers' balloon in 1783.

Smithsonian Institution

Fig. 1-5. Modern hot air balloons; these are quite similar to the very first balloons except that the heating device is carried along in the gondola.

Fig. 1-6. A small chunk of fluid that tends to remain static because all forces are in balance.

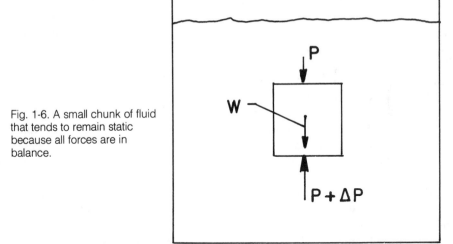

chunk of fluid had no weight, it would be pushed upward by this increased pressure on the bottom; however, the weight of this chunk of fluid added to the pressure force on the top surface balances out the increased pressure on the bottom and it remains still or *static*.

Suppose now that this chunk of fluid is replaced by a container filled with a fluid lighter than the surrounding fluid. In this case, the weight is not sufficient to balance out the increased pressure on the bottom and the container rises. Actually, the total weight of the container and the fluid inside it must be considered. This total weight must be less than the weight of the amount of fluid it displaces in order for it to rise.

This principle is what causes a balloon to ascend. Heated air or a lighter-than-air gas is placed inside the balloon. When the resulting weight of the balloon and the gas inside it is less than the surrounding air, the balloon ascends. The same thing happens to a cork placed in a depth of water. Being lighter than the water it displaces, the cork rises to the surface. A chunk of lead, which is heavier than the amount of water it displaces, sinks to the bottom.

Why then, you might wonder, does the balloon not rise to the very top of the atmosphere, as the cork rises in water? There is a very good reason for this. Water is essentially an incompressible fluid. The *density* of water—that is, its weight per unit of volume—does not change with depth. One cubic foot of water at the top of a tank weighs about 62.4 pounds (at standard sea level conditions) and one cubic foot several feet (or several hundred feet) down in the tank weighs the same. This is not true of air. Air is compressible and has no specific amount of weight associated with a certain volume (say, a cubic foot). The air compressed in an air compressor or in a pneumatic tire has a high weight per cubic foot, while that in the surrounding atmosphere has a much lower weight per cubic foot.

The atmosphere is a big "sea of air" that surrounds the Earth's surface just as a sea covers the ocean floor. The sea water is held there by the attraction of gravity; however, because the air is compressible, it is compressed by its own weight. Because all of the air is above the surface, the highest compression—and, hence, highest density—occurs at the surface. As you go up in altitude, there is less air above causing the compression at that altitude, and the air therefore becomes less dense (lower weight per cubic foot).

Now consider the balloon, which has a gas of a certain density inside, rising in the atmosphere. Eventually it will reach an elevation where the density of the air outside the balloon is the same as the air (or whatever gas) inside the balloon. At this point the forces are again in balance and the pressure difference between its top and bottom surfaces is not great enough to push the balloon upward (Fig. 1-7).

If the balloon were expandable, then the lighter gas inside would expand the balloon progressively as it rises, continuing the difference in densities between the inside and outside air. Such designs are made to allow balloon flights at very high altitudes. There is a limit, however, on the size to which any balloon can expand, and thus a corresponding limit on altitude. Such balloons must use a lighter-than-air gas, because temperature also decreases with altitude. Hot air would tend to cool down in very low outside temperatures, and is therefore not suitable for high-altitude flight.

Balloons are referred to as *aerostatic vehicles*, which means that they will lift in a static air mass, or one that is not moving. The balloon has no mechanism to move horizontally in this air mass. If the air mass itself is moving, then the balloon will move with it. Travel by balloon is thus pretty much of a chance operation because the route depends on the velocity and direction of the wind.

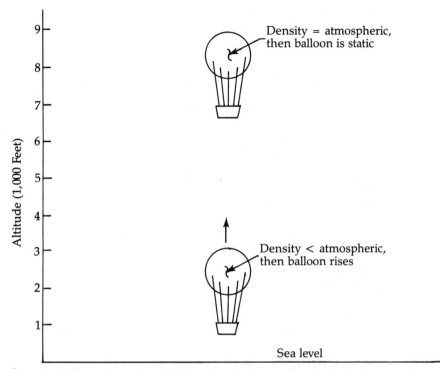

Fig. 1-7. A balloon whose density is less than the surrounding air rises until its density is equal to that of the surrounding air.

Early inventors attempted to make balloons somewhat more navigable by providing them with some propelling device. A lighter-than-air craft with propulsive capability is known as an *airship*. Early airships were quite slow due to the high drag of the balloon chamber. They were also somewhat cumbersome to maneuver. For a short time they enjoyed a degree of success as flying luxury liners in the period following World War I. Some thought is being given again today to using airships for special purposes because of their great fuel efficiency. Fuel is required only to provide propulsion and not lift, as it is with other types of aircraft.

However, early pioneers of flight recognized the limitations of lighter-than-air craft. They continued to observe the swiftness and maneuverability of birds and thought that somehow these qualities of flight could be achieved by humans.

## HEAVIER-THAN-AIR FLIGHT

In all of the years that men attempted to emulate the birds, they seemed to notice only one method of flying used by these creatures, namely wing flapping; however, there is another method that birds employ, some species more than others—that is, a gliding situation with their wings fixed in position. Many birds hold this fixed position for a considerable amount of time. Indeed, they even soar to higher altitudes in rising air currents without ever flapping their wings, except for an occasional course correction.

One of the first persons to recognize the fixed-wing mode as a possibility for manned flight was Sir George Cayley, an English nobleman who lived during the late 18th and early 19th centuries. He concentrated on developing the fixed-plane into a flying machine as avidly as da Vinci had pursued the flapping-wing concept.

Cayley was quite scientific in his approach and even used models on a whirling test rig to prove or disprove his theories. He was among the first to recognize the importance of stability and control in an aircraft and designed tail surfaces to achieve these qualities. His designs were quite similar to the basic conventional arrangement of wing and tail used in most airplanes today.

In 1853, Cayley designed and built a human-carrying glider. His coachman was pressed into service as the pilot and made the first manned flight of a heavier-than-air craft in history. Apparently the coachman was not as enamored with flight as Sir George, and promptly resigned after his first experience, never realizing his significance in aeronautical history.

Although Cayley's work was published, it never received wide recognition until very recently. Other scientists and inventors experimented with gliders, but pretty much due to independent ideas and motivation. Notable among such individuals was Otto Lilienthal of Germany. Lilienthal was an engineer who approached the idea of flight with scientific reasoning and analysis. He made a number of successful glider flights, but was killed in one attempt before he got around to his ultimate goal of powered flight. Lilienthal flying one of his typical designs is shown in Fig. 1-8.

In the United States, Octave Chanute, a Chicago engineer, was experimenting with gliders about the same time as Lilienthal. This period was in the late 1800s. Chanute was also scientific in his approach and spent great efforts at researching and collating all of the previous work that had been done in aeronautics. As a civil engineer, he contributed significantly to the structural soundness of aircraft in designs such as that in Fig. 1-9.

In all of these efforts, the objective was the design of a powered, human-carrying aircraft. Of course, one reason that powered efforts failed in the 1800s was the lack of a lightweight engine with sufficient power. There were many who attempted flight with various powerplants that did exist. The steam engine was the primary source of power in that era, and it is not too surprising that aircraft were designed with such means of power.

None of these were really successful, however. Another reason that such efforts failed was the fact that their inventors seemed to concentrate only on the one aspect of flight, namely a powerful engine. As early as 1894, Sir Hiram Maxim of England succeeded in briefly lifting into the air a huge machine propelled by a powerful steam engine. It lacked any means of control, however, and promptly crashed.

Fig. 1-8. Otto Lilienthal making one of his successful glider flights in Germany in the late 1800s.

Fig. 1-9. Octave Chanute experimenting with gliders in the United States.

Such efforts were typical of those who considered the problem of flight to be entirely one of developing enough lift and propelling force. They seemed to feel that once this aspect had been achieved, the aircraft could be steered around through the air about as easily as a boat in the water. In reality, flight in a three-dimensional medium is quite different. An aircraft has to be controllable in pitch and roll as well as in direction. It must also be stable, or have a means of making it stable, and the person controlling it must have knowledge and skill in doing so.

Two men who recognized all of the problems involved in flight were Wilbur and Orville Wright. Some historians tend to treat them lightly, classifying them as mere tinkerers who stumbled onto a flyable design. Others have claimed that their success was due to their capitalizing on the experiments of earlier pioneers. Certainly they did this, but this is a very legitimate and necessary part of technical development.

In reality, the Wright brothers were extremely talented and foresighted young men. Their procedure was extremely methodical and scientific in approach. They studied the successes and failures of others—in particular, Lilienthal and Chanute—and pinpointed the problem areas. The famous first flight at Kitty Hawk was not a single effort, but the culmination of four years of research at this desolate location.

They tackled one problem at a time: construction techniques, devising controls, teaching themselves to fly, and increasing the efficiency of design. Early experiments were performed with gliders. They even used tethered gliders with mechanical instrumentation to measure the actual forces created on the craft by the wind. When they found much previous work to be in error, they went back to their hometown of Dayton to construct a wind tunnel and used it to solve some of the mysteries of lift and drag.

Only after their gliders had been perfected to the point where they were completely reliable and predictable did they proceed to attempt the application of power. The concentration was then on a lightweight—yet sufficiently powerful—engine and an efficient propeller. On December 17, 1903, these painstaking efforts paid off as Orville lifted off the sands of Kitty Hawk in the world's first *sustained, controlled, and powered* flight as shown in Fig. 1-10.

## THE SCIENCE OF AERODYNAMICS

Gliders and airplanes are *heavier-than-air* machines. They cannot be buoyed up in a static air mass like balloons, because they are heavier than the amount of air that they displace. They depend instead on moving through the air so that the movement of air over their surfaces causes lift. The study of the behavior of moving air and the forces that it produces is referred to as *aerodynamics*, meaning, literally, "air in motion." It is this branch of science that must be employed to understand how airplanes fly, as opposed to *aerostatics*, meaning "air at rest," which explains balloon flight.

Aerodynamics is quite a complicated subject, much more so than aerostatics. The exact behavior of wings and airfoils was not really understood until many

Fig. 1-10. The first successful powered flight of the Wright brothers in 1903 at Kitty Hawk, North Carolina.

years after the first airplanes were flying. Scientists began around 1910 to look into the physical principles that explain winged flight. Frederick Lanchester in England, Ludwig Prandtl in Germany, and the Russian scientist Joukowsky developed much of the classical theory of flight around that time on which most modern aerodynamics is based. These men were scientists, remember, intent on explaining physical behavior. They were not particularly interested in developing aircraft.

Airplane development pretty much relied upon gradual improvement of tried and proven methods. Even early efforts of the National Advisory Committee for Aeronautics (NACA), the forerunner of NASA, were directed at trial-and-error testing. It was well into the 1920s before scientific theory and experimental testing were combined into a unified effort to improve the efficiency of flight.

Because aerodynamics is a complex science, most books on this subject are very deep in scientific theory and use complicated mathematical equations to describe the behavior of air in motion. We will proceed, however, to explain the aerodynamics of flight in a manner understandable to anyone at all familiar with airplanes and the very basic fundamentals of flying.

# 2

# Lift

THE UNIQUE QUALITY OF AN AIRCRAFT, as opposed to all other vehicles, is its ability to rise into the air. The force that opposes this action is the pull of *gravity*. Gravity acts on all bodies on or near the surface of the Earth. The resulting force that gravity causes on some mass of matter is referred to as its *weight*. In order to rise into the air, therefore, a force must be created that is equal to or greater than the weight and acting in the opposite direction. We call this force *lift*. We saw in the case of balloons that this lift force was created by an imbalance between pressure over the surface of the balloon and the weight of the balloon. With the balloon and its contents being lighter than the surrounding air, the imbalance resulted in a net force in the upward direction.

Airplanes, however, are heavier than the air that they displace, and thus must rely on other means to develop lift. The wing is the device employed to develop lift on an airplane and relies on forces created as it moves through the air. In order to sustain the airplane in level flight, the lift developed by the wing must just balance out the weight force as shown in Fig. 2-1. The aerodynamic forces (forces resulting from moving air) are much more complex than those created by the static air on the balloon. In order to understand how wing lift is created, we must first consider some of the basic principles of physics that are involved.

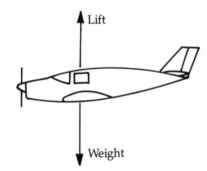

Fig. 2-1. Lift and weight must be in balance in sustained level flight.

# THE PHYSICS OF LIFT

Everyone involved in aviation surely must have heard of Bernoulli's equation. The principle discovered by this Swiss mathematician has been mentioned so often in the explanation of lift that many must feel he had something to do *directly* with the invention of the airplane.

Nothing could be further from the truth. Daniel Bernoulli never heard of an airplane. He died in 1782, the year before the Montgolfier brothers put the first balloon into the air. His work had absolutely nothing to do with flying at the time of his investigations. He was trying to explain, mathematically, the variation in pressure exerted by a moving mass of fluid, and was concerned mostly with flowing streams of water.

Mathematically, the equation that Bernoulli devised is usually written:

$$p + \frac{1}{2}\varrho V^2 = \text{constant}$$

The Greek symbol $\varrho$ (pronounced "roh") is commonly used to designate the air density—that is, the mass (or weight) per cubic foot. The other symbols, p and V, of course, indicate pressure and velocity.

In plainer language, this equation says that the pressure plus 1/2 times the density times the velocity squared must always equal a constant value (in an open, continuous flow of fluid), or:

$$\text{pressure} + \frac{1}{2} \times \text{density} \times \text{velocity squared} = \text{constant}$$

There is another way to consider this equation, which makes it even simpler. The term $1/2\, \varrho\, V^2$ is also referred to, collectively, as the *dynamic pressure*. The p in the equation is really the static air pressure. Consequently, we can state the equation as:

$$\text{static pressure} + \text{dynamic pressure} = \text{constant}$$

Static again means "still" and is used to describe the pressure exerted by a static, or still, air mass. A mass of stationary air (or any fluid) in a container would exert a certain amount of static pressure on the surrounding walls of the container, as shown in Fig. 2-2. Now, if the air were in motion, it would still have some *static pressure* associated with it, which would be exerted on the walls of the container through which it is flowing, as shown in Fig. 2-3; however, the air would also have a *dynamic pressure* associated with it. Dynamic pressure is the pressure that would be exerted if the fluid were brought to rest. This pressure would be realized if, for example, a plate were placed in the fluid in an effort to dam the flow, as also shown in Fig. 2-3.

Now, the point of Bernoulli's principle is that the fluid only contains so much pressure. If the fluid is at rest, all of the pressure is static pressure, but if the fluid

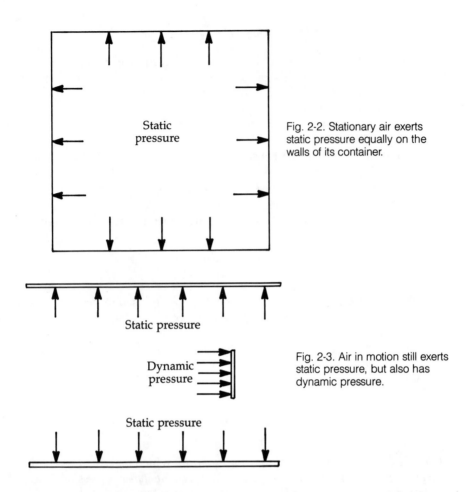

Static
pressure

Fig. 2-2. Stationary air exerts
static pressure equally on the
walls of its container.

Static pressure

Dynamic
pressure

Fig. 2-3. Air in motion still exerts
static pressure, but also has
dynamic pressure.

Static pressure

is in motion, then some of the static pressure must be traded off for dynamic pressure; therefore, as the flow speeds up, there is an increase in dynamic pressure, and the static pressure must be reduced because the sum of the two pressures must always be the same. This situation is shown in Fig. 2-4. Suppose, for example, that the static pressure on the left side of the figure had a value of 6 units and the dynamic pressure had the same value. The total constant pressure, then, would be 12, or:

static pressure + dynamic pressure = total (constant)
6 + 6 = 12

Now suppose that the fluid is speeded up on the right side of the figure so that the dynamic pressure is now 8 units. Because the static plus the dynamic pressure must still equal the constant total value of 12, the static pressure must go down to 4:

4 + 8 = 12

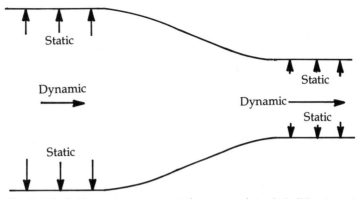

Fig. 2-4. As fluid speeds up, some static pressure is traded off for dynamic pressure.

This principle of science has been well proven and is used in quite a number of applications other than aeronautics. It simply relates static and dynamic pressures, and because dynamic pressure depends on velocity, can be thought of as relating static pressure and velocity. Notice, however, that Bernoulli's equation does not contain any explanation of why a fluid should speed up or slow down. It merely states that if it does, then the pressure will vary in the opposite way.

## Continuity

To explain why flows change velocity, we must resort to another very important principle of science, rarely spoken of in the intimate confines of the cockpit. This principle is known as the "Law of Continuity" or simply, the *continuity equation*. This law is really the *Law of Conservation of Matter* applied to a moving fluid. It states that in a moving stream of fluid, the density times the cross sectional area of the flow times the velocity must always be a constant or,

$$\varrho \times A \times V = \text{constant}$$

For flows at velocities much less than the speed of sound, the density will not vary in an open stream; thus, the density could be deleted from the equation and it could be stated as, simply, area times velocity equals a constant.

$$A \times V = \text{constant}$$

## Venturi tubes

This principle is most easily illustrated by a venturi tube, which is a tube with a restricted area of flow. Such tubes were used rather effectively to power the gyro instruments in older airplanes before the days of vacuum pumps. Figure 2-5 is a diagram of a venturi tube with the area reduced by $1/2$. In the wide part, the

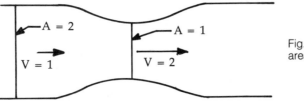

Fig. 2-5. A venturi tube with an area reduction of one half.

area is 2 units and the velocity 1 unit. Where the area necks down to 1 unit, the velocity must increase to 2 units so that:

$$1 \times 2 = 2 = \text{constant}$$

This principle can also be observed in a river or other stream of water. Where the banks of the stream narrow down, the flow speeds up; when the banks widen, the stream slows down.

The application of the venturi tube in creating a lowered pressure—or *suction*—can now easily be explained. In Fig. 2-6, a tube is shown connected to the throat of the tube. A flow coming into the tube will have an increased velocity, due to the reduced area, by the continuity law. If the velocity is increased, the pressure will be reduced, by Bernoulli's principle; hence, a tube connected at this point will experience this decreased pressure and thus create a suction on wherever it is connected. This is how vacuum (or suction) is created to drive the vacuum gyros with this type of gadget.

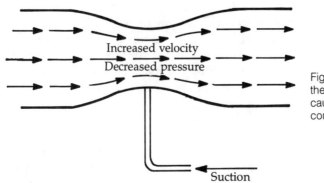

Increased velocity
Decreased pressure

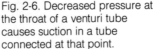

Suction

Fig. 2-6. Decreased pressure at the throat of a venturi tube causes suction in a tube connected at that point.

This device has many other applications. It is the method by which paint and garden sprayers work, it is the way fuel is drawn into a carburetor throat, and it is the principle on which vacuum cleaners operate. Even a lady preparing for her evening social activities by spraying perfume from an atomizer is making use of this very important principle of science.

## Other aerodynamic devices

The reduction in static pressure with increased dynamic pressure can be demonstrated by many very simple devices. One of the classic methods involves a thread spool as shown in Fig. 2-7. A small piece of cardboard (such as from a tablet back) is cut and a straight pin stuck through its center. The spool is held vertically and the cardboard held below it with the pin inserted into the lower portion of the hole through the spool. By blowing through the hole in the spool from the top, a jet of air is ejected radially from the bottom of the hole, spreading out over the top of the cardboard. With sufficient velocity, the pressure will be lowered enough on the top of the cardboard so that the higher pressure in the atmosphere will hold it close to the spool. You can now let go of the cardboard and it will stay in place, the pressure force overcoming the weight of the cardboard.

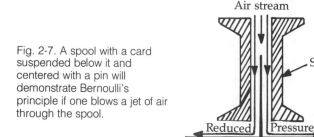

Fig. 2-7. A spool with a card suspended below it and centered with a pin will demonstrate Bernoulli's principle if one blows a jet of air through the spool.

Another even simpler device is to bend a small sheet of paper around your fingers in front of you as shown in Fig. 2-8A. By blowing over the top of the sheet, the pressure is lowered and the higher pressure below raises it up, as seen in Fig. 2- 8B. More dramatic demonstrations involve a jet of air from a compressed air source.

One way to utilize an air jet for such purposes is to blow it between two ping-pong balls suspended near each other. Rather than blow the balls apart, as one might reason, the lower pressure in the jet causes the balls to be drawn together. This setup is shown in Fig. 2-9. Probably the most amazing demonstration is to shoot the jet vertically upward and place a ping-pong ball in the jet airstream. It will remain there. It will rise until the force of the jet on the bottom equals its weight, but it will not be blown away, as might be imagined. Instead, the lower pressure holds the ball right in the center of the jet, as shown in Fig. 2-10. Any attempt to move it slightly to the side will result in the ball being drawn directly back into the center when it is released.

Such displays demonstrate some of the subtler principles of nature that are not intuitively reasoned. Air in motion displays many such subtle characteristics and is one reason why many of the true principles of flight are not clearly understood and why so many fables exist.

## How airfoils create lift

Now let us see what all of this science lesson has to do with airplane flight. First of all, we will consider just the two-dimensional effects on a profile cross

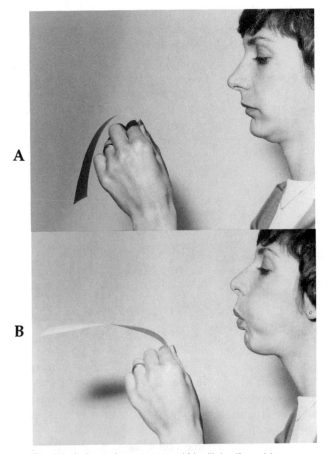

Fig. 2-8. A sheet of paper at rest (A) will rise if one blows over it, thereby decreasing the pressure on the top surface (B).

section of the wing. This typical cross-sectional shape is referred to as an *airfoil*. Picture, if you will, a very large airfoil moving through the atmosphere, as in Fig. 2-11. The upper boundary would represent the upper extent of the atmosphere and the lower boundary, the Earth's surface. Another way to view this situation is to consider this airfoil as being in a wind tunnel with the upper and lower boundaries being the ceiling and floor of the wind tunnel. (Wind tunnels will be discussed in detail later. They are used to simulate flight conditions for testing purposes. It makes no difference in the aerodynamic forces created whether the airfoil is moving through the air (which is stationary) or whether the air is moving over a stationary airfoil. The point being made here is simply that a relative motion exists between the airfoil and the surrounding air and that certain dimensional limits exist in the stream of air.)

Far ahead of the airfoil, the air has a uniform velocity associated with it across the entire stream. When it reaches the airfoil, however, it must divide, with part of the stream flowing over the top surface and part over the bottom.

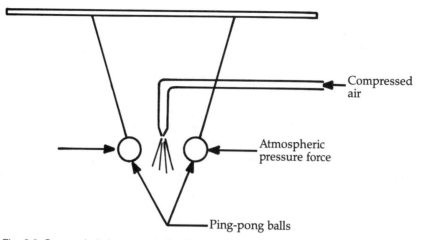

Fig. 2-9. Suspended ping-pong balls will be drawn together by the lowered pressure of a jet of air blown between them.

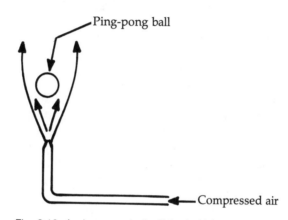

Fig. 2-10. A ping-pong ball will be held in an upward-directed jet of air due to the lowered pressure.

The cross- sectional area of the stream is reduced here, however, by the amount of space taken up by the airfoil. According to our continuity law, this reduction in area must cause an increase in velocity. As the thickness increases proceeding over the airfoil, the stream area decreases and hence its velocity increases proportionately.

Now, if we recall the Bernoulli principle, an increase in velocity (or dynamic pressure) dictates a decrease in static pressure; thus, by simply inserting the airfoil in the moving flow, we have caused the following effects on the airstream:

lower area → higher velocity → lower pressure

Notice that some of the stream goes over both the top and bottom surfaces. This effect is felt on both top and bottom; however, with an airfoil of typical

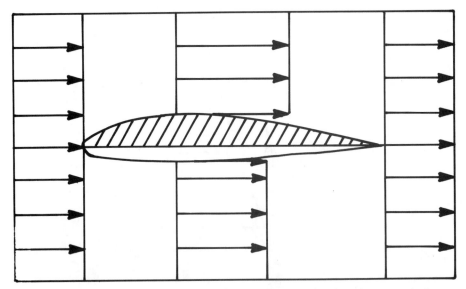

Fig. 2-11. An airfoil speeds up the air flowing over it proportional to the amount of area obstructed by the upper and lower portions of the airfoil.

shape, the area of the stream is reduced more over the top surface. The end result is a greater lowering of pressure on the top surface than on the bottom, even though the pressure is lowered somewhat over the bottom surface (from that of the free stream ahead of and behind the airfoil). The result is an imbalance of pressure forces between the top and bottom of the airfoil (Fig. 2-12) and we have a net force in the upward direction, which we term *lift*. The result is the same as the imbalance of pressure forces on the balloon, except that this lift has been created on a heavier- than-air body by the motion of the air over it, or by *aerodynamics*.

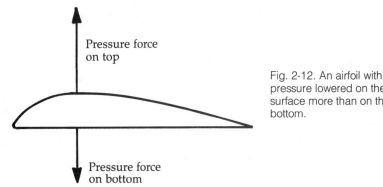

Pressure force on top

Pressure force on bottom

Fig. 2-12. An airfoil with pressure lowered on the top surface more than on the bottom.

## Airfoil terminology

Before proceeding with this discussion, we should discuss some terminology and dimensions pertaining to airfoils. Figure 2-13 shows a typical airfoil. The

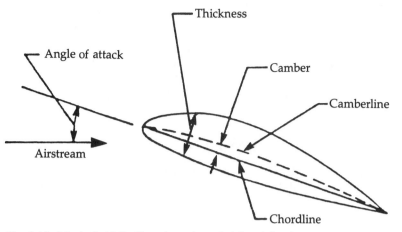

Fig. 2-13. A typical airfoil with various characteristics defined.

straight line connecting the *leading edge*, or forward most tip, to the *trailing edge* is called the *chordline* and the distance between the leading and trailing edges is referred to as the *chord*. The *angle of attack* is always measured between the chordline and the relative wind. The Greek letter $\alpha$ (alpha) is frequently used to denote this angle.

Note that for the typical airfoil shape shown, there is more cross-sectional area above the chordline than below. In other words, it is not symmetrical about the chordline. The line connecting all of the points midway between the upper and lower surfaces is called the *meanline*. This line is also sometimes referred to as the *midline* or *mean camber line*. The perpendicular distance between the chordline and the meanline is called the *camber*. There is some point along the chord where this distance is greatest, and hence at this location the airfoil has its maximum camber. The *maximum* camber of an airfoil is sometimes referred to simply as "the camber" because it is the only camber dimension of significance. The amount of camber, then, is a measure of the overall curvature of the airfoil. The distance between the upper and lower surfaces is called, simply, the *thickness*.

Cambered airfoils are normally used to more effectively provide lift in a given direction. This direction is usually upward on a conventional wing. In some applications, however, lift is sometimes needed in both directions, as with tail surfaces. Aerobatic airplanes also sometimes fly inverted, and thus must develop lift in a direction opposite to that in normal flight. For such applications a special airfoil is used, known as a *symmetrical airfoil* shown in Fig. 2-14.

A symmetrical airfoil has no camber and, in this case, the meanline and the chordline coincide. All airfoils are thus either symmetrical or cambered. It should be noted that in some aeronautical circles, the curvature of the upper surface of an airfoil is called "upper camber" and the curvature of the lower surface is referred to as "lower camber." This usage is not strictly correct because a symmetrical airfoil has curvature on its upper and lower surfaces, yet it has no camber in the true meaning of the term.

Fig. 2-14. A symmetrical airfoil with a coinciding chordline and meanline.

## Lift on cambered airfoils

Returning to our discussion of lift production, it should now be apparent that the "typical" airfoil that we have been discussing is really a cambered airfoil. In the example in Fig. 2-11, the airfoil is at zero angle of attack. A cambered airfoil will produce some lift at this angle because there is more cross-sectional area above the chordline than below, causing a greater reduction in the area available for the airflow (Fig. 2-11). At this angle of attack, the flow will divide very near to the leading edge.

Now let us see what happens when the angle of attack is increased. Figure 2-15 shows the cambered airfoil at some small positive angle of attack. In this picture, a very important principle of aerodynamics (but one also very obscure among the hangar crowd) becomes apparent. The flow no longer divides precisely at the tip of the leading edge, but at a point farther down on the nose of the airfoil. The dividing point is called the *stagnation point*. It is so named because the streamline (or line followed by one row of flowing air particles) completely stops, or *stagnates*, at this point. All of the streamlines above this point flow over the top surface and all of those below it go over the bottom.

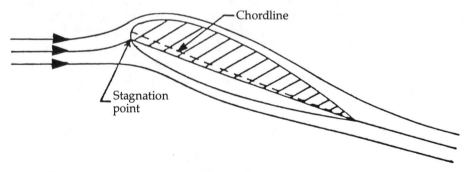

Fig. 2-15. A cambered airfoil at a moderate angle of attack showing the increased effective upper area due to the location of the stagnation point.

The *effective* upper cross-sectional area of the airfoil is therefore increased and the effective lower area is decreased. What happens is that the area of the airflow over the top surface is reduced even more than at lower angles of attack. Again, from our consideration of continuity and the Bernoulli principle:

lower area → higher velocity → lower pressure

Hence, a greater effective upper surface area leads to a greater lowering of pressure on the top surface. The reverse is also true on the lower surface. The reduction in effective cross-sectional area here has reduced the airflow area and resulted in *less* lowering of pressure on the bottom surface. The result, as shown in Fig. 2-16, is a greater resulting pressure force in the upward direction, or greater lift.

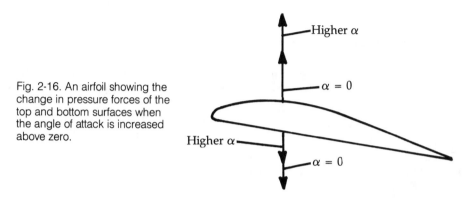

Fig. 2-16. An airfoil showing the change in pressure forces of the top and bottom surfaces when the angle of attack is increased above zero.

As the angle of attack gets even greater, the stagnation point moves farther down on the airfoil. The effect is a greater lowering of pressure on the top surface than on the bottom, resulting in even greater lift. This discussion explains the reason that increased angle of attack causes increased lift, a fact that every pilot should be aware of.

It should now be obvious how lift can also be created by a symmetrical airfoil. At zero angle of attack, the stagnation point is directly on the leading edge and the flow over the top and bottom surfaces is identical. Pressure is reduced over both surfaces equally, and although there is both an upward and a downward pressure force created, they are equal, and the net effect is zero. This situation is shown in Fig. 2-17.

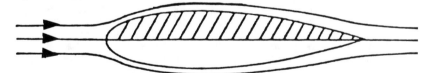

Fig. 2-17. A symmetrical airfoil at zero angle of attack with equal upper and lower surface area.

## Symmetrical and inverted airfoils

Now, if we put the symmetrical airfoil at an angle of attack as shown in Fig. 2-18, the stagnation point moves below the leading edge point. The effective upper and lower cross-sectional areas are then different, and the same effects occur as with the cambered airfoil. Because it is symmetrical, however, a greater angle of attack will be required to get the same amount of lift as with a cambered

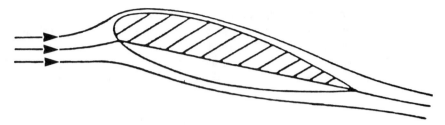

Fig. 2-18. A symmetrical airfoil with increased effective upper surface area at higher angle of attack due to lowered stagnation point.

airfoil; hence, the symmetrical airfoil is not as efficient in this respect. Its advantage is the fact that it can produce an equal amount of lift in either direction at the same positive or negative angle of attack.

We can also produce negative lift with a cambered airfoil, but a very great negative angle is required. This means that it is possible to fly inverted with a cambered airfoil. The inverted angle must be great enough, though, that the effective lower area of the airfoil (which is now, in reality, the upper) must be greater than the upper (now the lower), as seen in Fig. 2-19. Any pilot who has flown (or attempted to fly) a cambered wing in inverted flight is aware of this high angle required and the accompanying forward stick position that must be maintained.

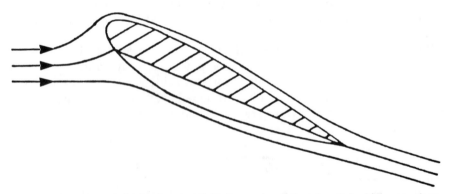

Fig. 2-19. A cambered airfoil in inverted flight; the angle of attack has to be high enough so that the lower surface area (now directed upward) is greater than the upper area.

## PRESSURE DISTRIBUTION
## AND PITCHING MOMENT

So far we have been talking about pressure on the top surface or the bottom surface as if it were constant over these surfaces. Such is not the case. The pressure is actually different at each point along the chord. Remember also that the significance of the pressure along these surfaces is the *difference* of this pressure from that of the surrounding atmosphere. Anywhere that we have air, it will have certain pressure associated with it. If we can create a pressure lower or higher than the surrounding air (or *ambient air*, as it is called), we create an aerodynamic force by virtue of the pressure difference (sometimes referred to as *differential pressure*).

Figure 2-20 shows a cambered airfoil at an angle of attack where it develops zero lift. Notice that this angle has to be slightly negative because the cambered airfoil produces some lift at an angle of attack of zero. The arrows indicate the amount and direction of pressure forces at various points along the airfoil surface. Arrows pointing away from the surface indicate lower pressure than the surrounding air; hence, a force pulling away from the surface, also referred to as a *negative pressure*. Arrows pointing toward the surface indicate higher pressure at these points, meaning that the surrounding air would exert a force against (or push on) the surface, also called *positive pressure*. In this zero-lift condition, the sum of all the pressure forces tending to pull upward exactly equals those tending to pull downward and a net force of zero results.

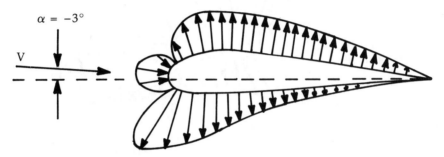

Fig. 2-20. Pressure distribution on an airfoil at the angle for zero lift ($-3°$).

Now if we raise this same airfoil to a slightly positive angle of attack, the pressure distribution results in that shown in Fig. 2-21. Here we can see that there is more negative pressure in the upward direction than in the downward, and we would get some positive lift. This angle of attack would be typical of a cruising condition. Notice that there is still a reduced pressure on the bottom surface, but the net effect is an upward force and this force would really be provided by the higher pressure of the surrounding air.

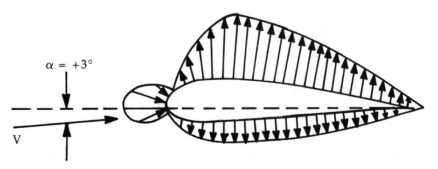

Fig. 2-21. Pressure distribution on an airfoil at a slight positive angle of attack typical of normal cruising flight.

Finally, in Fig. 2-22 we see the airfoil at a very high angle of attack. The pressure on the top surface is very much reduced. Notice also that now we have positive pressure over most of the lower surface, adding to the lift. The uninformed observer of airplanes in flight probably has the idea that this positive pressure against the lower surface is the sole creator of lift. In reality, it is only a small part. It is this picture, undoubtedly, that has also inspired the old hangar-tale of so much of the lift coming from the top surface and so much from the bottom.

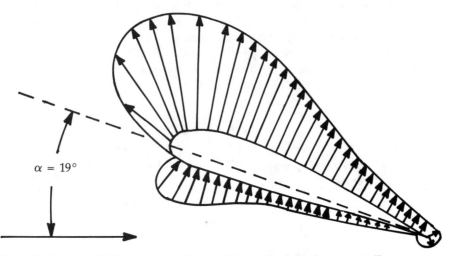

$\alpha = 19°$

Fig. 2-22. Pressure distribution on an airfoil at a high angle of attack near to stalling.

The pressure distribution changes with every angle of attack, as these figures have shown. At cruising angles there is even negative pressure on the bottom surface. The lift comes from the *net* effect. It would be very difficult to separate the effects of the top and bottom surfaces because the pressure distribution on both surfaces depends on the overall shape of the airfoil.

Consider now the chordwise variation in pressure. If we look at the airfoil at 3° angle of attack in Fig. 2-21, as one proceeds back from the leading edge, the pressure gets lower. This is because the airfoil gets thicker and is interrupting more of the airflow. At about its maximum thickness point, the negative pressure reaches a maximum (or minimum, depending on how you look at it), and then drops off again. The negative pressure continues to decrease as the airfoil gets narrower until finally it goes to zero at the trailing edge. At this point the airfoil thickness also becomes zero. Here the pressure now becomes equal to that of the surrounding atmosphere. Remember that the negative pressure was really a *differential* from atmospheric, so that when the *difference* goes to zero, the pressure then becomes atmospheric.

If we were actually to pull up on the airfoil with a force proportional to the length of each arrow at the point of each of these arrows, we would simulate the total pressure force. We could also simulate this pressure distribution by a *single* force applied at the centroid of this distribution, known as the *center of pressure*. It is analogous to the center of gravity when considering the gravity forces (or weight) distributed over a body. The center of pressure varies with angle of attack so that the resultant force acts at a different point on the airfoil for each angle of attack.

We have seen that the pressure distribution is not the same on both top and bottom surfaces. This leads to the fact that the center of pressure could also be different for top and bottom surfaces. In most cases there is a difference, and usually the lower surface center of pressure is forward of that of the upper surface. The resulting situation is shown in Fig. 2-23. Because the resultant forces are not acting at the same chordwise location, a rotation of the airfoil will result. A rotating or twisting tendency is called a *moment* or, sometimes, *torque*. A moment can be defined as a force times a moment arm.

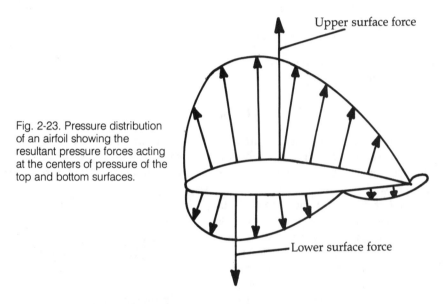

Upper surface force

Fig. 2-23. Pressure distribution of an airfoil showing the resultant pressure forces acting at the centers of pressure of the top and bottom surfaces.

Lower surface force

Either an increase in force or a lengthening of the distance (arm) at which the force is applied will increase the moment; thus, the farther apart the centers of pressure of the upper and lower surfaces are, the greater will be the resulting moment. This specific moment is referred to as a *pitching moment* because rotation about the lateral or spanwise axis is called *pitch*. This particular consideration of the airfoil shows that it is, by itself, unstable, and requires some additional mechanism to stabilize it before it can fly. The lack of understanding of this principle led to the untimely demise of many early pioneers of flight.

With changes in angle of attack, both the lift force and the center of pressure will change. There is a point on the airfoil, however, about which the pitching moment does not change, but remains constant with all angles of attack. This point is called the *aerodynamic center* and is located at approximately the quarter-chord position (25 percent of the chord length aft of the leading edge) for most airfoils. It is quite convenient to always consider the lift force to act at this point with a pitching moment of constant strength acting to provide a rotational effect about the point, shown in Fig. 2-24. This method represents the total effect of lift

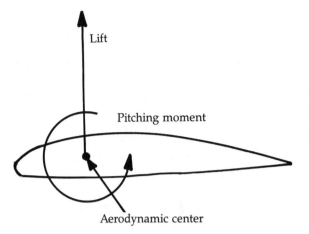

Lift

Pitching moment

Aerodynamic center

Fig. 2-24. Representation of pressure force by a single lift vector at the aerodynamic center with a pitching moment about it.

in a simple—yet accurate—way and is commonly used by most all of those involved in aerodynamics.

## THE STALL

We have seen that as the angle of attack increases on an airfoil, the lift increases. This increase can't go on indefinitely and intuitively one might reason that there would be a practical limit to angle of attack. There is indeed such a limit, and this limit is known as the *stalling point*.

Several things contribute to the stall. First of all, as the angle of attack increases, the stagnation point moves farther down on the forward part of the airfoil, making a longer effective upper surface. The flow over the top has a longer path to travel. As the air flows over a surface, a certain amount of friction is created. (This phenomenon is discussed in greater detail in chapter 3.) As the path gets longer, the frictional force gets greater.

There is also an effect resulting from what is known as *pressure gradient*. *Gradient* is a word meaning rate of change. Initially, proceeding back from the leading edge, the pressure decreases with distance. The decreasing pressure tends to induce the flow to move along the surface, and hence this decreasing tendency in the pressure is known as a *favorable pressure gradient*. That is, it is favorable to promoting the flow in the direction we want it to go. Beyond the peak in the negative pressure, we have a reversal in pressure. It now increases with distance or becomes less negative. This condition works against the flow and is known as an *unfavorable pressure gradient*. As the angle of attack increases, the center of pressure moves forward and the unfavorable pressure gradient becomes longer and steeper.

Eventually, the combined effect of the unfavorable pressure gradient and the surface friction becomes greater than the energy available in the airflow to overcome them. At this point the flow will detach itself from the surface, or *separate* as the term is sometimes used. This condition is shown in Fig. 2-25. With no flow over the top surface, there is no longer a mechanism to reduce the pressure over

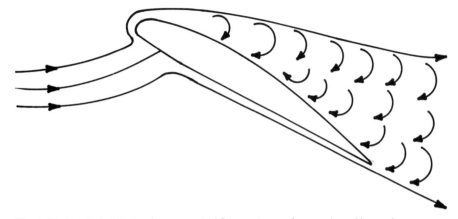

Fig. 2-25. Airfoil at stall showing separated flow on top surface and resulting wake.

the surface and lift decreases drastically. A wake of random, turbulent air now exists over this surface, contributing practically nothing to lift. This phenomenon is known as a *stall*.

The lift does not really go to zero because there is still flow over the lower surface and at this angle of attack is normally exerting positive pressure. The upper surface separation causes a great loss in lift production. The result on an aircraft in flight is a sudden loss of necessary lift; it will drop due to weight now being greater than lift.

It should be noted that there is a definite angle of attack associated with the stall for each airfoil shape. A plot of lift versus angle of attack for a typical airfoil as shown in Fig. 2-26. The lift will increase in proportion to angle of attack and then drop off sharply at the stall angle, reaching a maximum just before the stall. Exactly how rapidly this occurs depends on the airfoil shape.

One factor that reduces the abruptness of the stall is the roundness of the leading edge. In fact, a very sharp leading edge can act almost as a barrier to the flow at high angles of attack and cause the flow to separate right there. Sometimes it is desirable to induce a stall to occur at one certain spanwise wing location before another. In this case, a very sharp leading edge device known as a *stall strip* is installed, which causes the flow to separate at the leading edge at an angle of attack somewhat below the normal stall angle.

Because the stalling of an airplane wing is not the most desirable situation—particularly when it occurs near the ground—a number of devices have been designed to warn the pilot of an impending stall. The most popular of these is the *vane-type stall warner*, illustrated in Fig. 2-27, which takes advantage of the relation between stall angle of attack and stagnation point. Remember that the stagnation point moves down along the nose of the airfoil as the angle of attack increases. There is a distinct stagnation point (point of division in upper and lower surface flow) for each angle of attack.

The vane is positioned so that the stagnation point is above it in normal flight as in Fig. 2-27A. The airstream hitting the vane is, then, that going over the lower

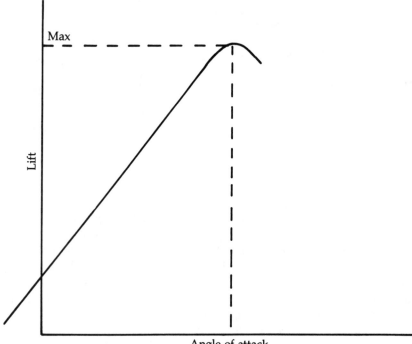

Fig. 2-26. Plot of lift versus angle of attack showing maximum lift at the stall angle.

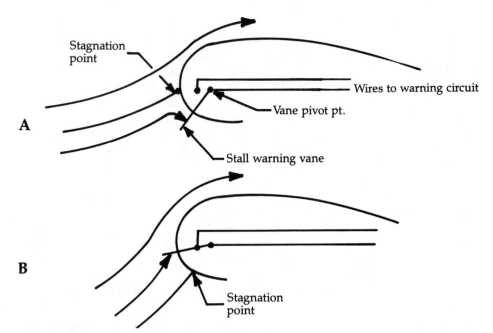

Fig. 2-27. Stall warning device with stagnation point above vane (A) and near stall condition with stagnation point below vane (B).

surface, which tends to hold the vane down. The vane is connected to an electrical switch, which is open when the vane is down. As the angle of attack is increased, the stagnation point moves downward. As the stall angle is approached, it moves below the vane. The airstream is now moving toward the upper surface in the vicinity of the vane, as shown in Fig. 2-27B. This flow pushes the vane up, closing the switch. The switch is connected to an electrical circuit that activates a warning horn or light to warn the pilot. By adjusting the switch slightly up or down on the nose, the margin of warning prior to the actual occurrence of stall can be altered.

# AIRFOIL DEVELOPMENT AND DESIGNATION

We have been talking a lot about the "typical" airfoil shape. Actually, there are many different airfoil shapes. Slight variations in shape alter their characteristics in one way or another. At this point it would be appropriate to consider some of these characteristics and to see how the "typical" airfoil shape came into being.

Almost any relatively flat surface could be made to generate some lift. In fact, a perfectly flat thin plate will do the job. If you don't believe that, try out any of a number of simple little balsa wood hand-launched model gliders. Most of them have flat wing sections, and they fly. The flat plate, then, is probably the simplest of airfoil sections, as shown in Fig. 2-28A. At zero angle of attack, it produces no lift because it is actually a symmetrical airfoil (it has no camber); however, at a slightly positive angle, it will produce lift, as shown in Fig. 2-28B.

Notice that at this angle, the stagnation point is not on the leading edge, just as with any airfoil, but farther back along the lower surface. The flow division here gives a longer path over the top surface and a lowered pressure over the top surface exists; therefore, even the flat plate performs the normal airfoil function of creating a pressure differential. It does not lift simply by diverting the airflow downward with the bottom surface as, for example, water skis do.

The flat plate is not a very efficient airfoil because it creates a fair amount of drag, as will be discussed later. The sharp leading edge also promotes stall at a very small angle of attack, and thereby severely limits its lift-producing ability. This situation is seen in Fig. 2-28C.

Whether or not anyone ever actually flew an early glider with a completely flat airfoil is not too well documented; however, there are many records of the use of airfoil shapes similar to that shown in Fig. 2-29A. Note that this shape is merely a flat plate with the nose bent downward. This modification enabled the airfoil to achieve much higher angles of attack without stalling; however, it was efficient only over a very small range of angles. At both higher and lower angles, the sharp leading edge still induced separation. Figure 2-29B shows separation on the lower surface at low angles of attack. It must have occurred to someone, then, to fill in this area and yield a somewhat rounded nose so that a wider range of angles of attack was possible, resulting in a shape somewhat like that of Fig. 2-29C.

Airfoils of this general shape were used for many years in early gliders and airplanes. Most of them were devised by simply testing a wide variety of varia-

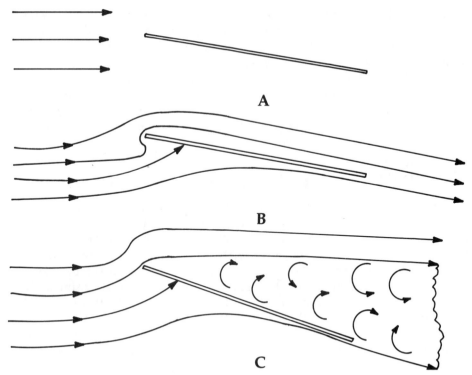

Fig. 2-28. A flat plate airfoil (A) showing flow at a slight angle of attack (B) and during a stalled condition (C).

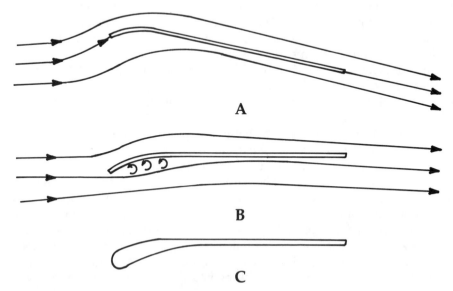

Fig. 2-29. A curved thin airfoil showing stagnation point right on leading edge (A), stagnation point above leading edge (B), and thickened leading edge modification (C).

tions in basic design. The Wright brothers were quite active in this area. In a two-month period in 1901, they tested more than 200 different airfoil shapes in their homemade wind tunnel. It should be obvious also that this basic shape is typical of the airfoil of a bird's wing. (In all probability, the original idea came from that source.)

Very early airplanes relied heavily on external bracing with struts and wires to provide necessary strength. The thickness of an airfoil was not too important from this consideration. As structural design developed, though, it became apparent that internal support by spars greatly reduced the drag of wings. In order to provide depth for the spar, greater thickness had to be added to the airfoil. These thicker airfoils also displayed greater lifting capability and finally evolved into the shape in Fig. 2-30, which we recognize today as the "typical" airfoil.

Fig. 2-30. Typical thick airfoil shape that has evolved to provide for internal structure.

Until quite recently, airfoil design was far from an exact science. It was instead an evolutionary process that relied totally on testing in flight and in wind tunnels. In the very early days there was no orderly system of identifying airfoils. Those that seemed to prove effective were given designations, but simply in some arbitrary order, usually of chronological development. The RAF 6, the Gottingen G-398, and the Clark Y are typical airfoil examples. Because only slight variations in airfoil design make large differences in aerodynamic performance, a large number of different airfoils came into being by approximately 1920. It was obvious that some orderly system of identifying airfoils had to be devised.

The National Advisory Committee for Aeronautics (NACA), which was the forerunner of NASA, undertook this task in the late 1920s. Their wind tunnel tests showed that the aerodynamic characteristics of airfoils depend primarily upon two shape variables: the thickness form and the meanline form. They then proceeded to identify these characteristics in the numbering system for the airfoils.

The first such airfoils are referred to as the NACA four-digit series. The NACA 2412 (pronounced "twenty-four twelve") airfoil is a typical example, shown in Fig. 2-31. The first number (2 in this case) is the maximum camber in percent (or hundredths) of chord length. The second number, 4, represents the location of the maximum camber point in tenths of chord and the last two numbers, 12, identify the maximum thickness in percent of chord. All characteristics are based on chord length (c) because they are all proportional to the chord. For this airfoil, the maximum camber is 0.02c, the location of maximum camber is 0.4c, and the maximum thickness is 0.12c; thus, if one were to use this airfoil in a wing with a 60-inch chord, the maximum camber would be 1.2 inches, located 24 inches behind the leading edge, and the maximum thickness would be 7.2 inches.

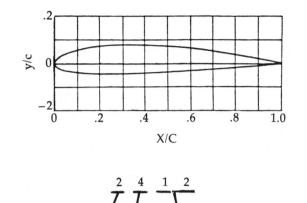

Fig. 2-31. Meaning of numbers in a typical four-digit airfoil (NACA 2412).

Such numbering systems led to families of airfoils. For example, a 2415 airfoil would have the same meanline shape as the 2412, but would be 3 percent thicker. The 60-inch chord airfoil mentioned above would then have a maximum thickness of 9 inches with the 2415 shape, while other dimensions remained the same. Many of its characteristics would be similar to those of the 2412, with only those affected by thickness being altered. This is an important consideration if one wants the basic characteristics of a certain airfoil but needs a little more thickness to provide for a spar of adequate strength.

Four-digit airfoils that have no camber, or are symmetrical, would have two zeros in the first two digits. A 0010, for example, is a symmetrical airfoil frequently used in tail surfaces and is commonly pronounced "double-oh ten" (not to be confused with a British secret agent).

The meanline basic shape was altered slightly in an improved airfoil developed in the mid 1930s. The position of maximum camber was moved farther forward and resulted in 10 to 20 percent greater possible maximum lift. To properly identify these airfoils, NACA developed the five-digit series. The NACA 23012 is an example of such an airfoil and is used on the Beech Bonanza. The first and last two numbers designate camber and thickness pretty much as the four-digit series do. The only difference is that the position of maximum camber (the 3 here) indicates twentieths of chord, rather then tenths as in the four-digit. The major difference is the addition of the fifth digit in the middle. A zero (as used in 23012) indicates a straight line aft meanline and a 1 here would indicate a curved aft meanline.

In the late 1930s, as aeronautical technology became more sophisticated, a number of new series of airfoils were developed. Most of these efforts were aimed at airfoils capable of speeds in the 300 to 400 mph range, which were then coming into being. One of the basic concepts was to design airfoils with a minimum (most negative) pressure farther back on the airfoil to provide a greater range of favorable pressure gradient. Another consideration was to optimize a particular airfoil (make it most efficient) at a certain angle of attack or *lift coefficient*. Lift coefficient is a measure of the lift at a certain angle of attack and is discussed elsewhere in this chapter.

The first such airfoils were known as the 1-series, typified by airfoils such as the 16-218. The most popular series is the 6-series and these are used quite extensively today on many lightplanes, including Cherokees and Mooneys. An example is the NACA $65_2$-415, depicted in Fig. 2-32. The 6 designates the series. The second number, 5, is the location, in tenths of chord, of the minimum pressure point. The subscript, 2, indicates the range of lift coefficients above and below the design lift coefficient where low drag can be maintained, in this case, 0.2. The next number, 4, indicates the design lift coefficient of 0.4, and the last two digits, 12, again mean maximum thickness in percent of chord.

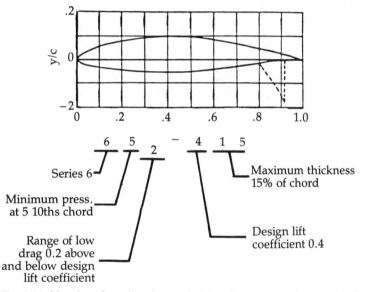

Fig. 2-32. Meaning of numbers in a typical 6-series airfoil (NACA, $65_2 - 415$).

The 6-series airfoils were first used in the wing of the P-51 Mustang for their low-drag qualities. Their popularity today stems as much from their convenient shape. In order to achieve a far aft minimum pressure point, the maximum thickness also results at a pretty far aft location. This enables a maximum depth spar to be located at this aft position and turns out, very conveniently, to be very nearly directly beneath the rear seat in a four-place light airplane. Of course, the sales departments of most companies are also quite enthusiastic about being able to boast that their airplanes have modern, low-drag airfoils similar to World War II fighters.

The next major breakthrough in airfoil development came in the later 1960s when Richard Whitcomb, a NASA research engineer, devised the supercritical airfoil. This airfoil was intended to improve drag at speeds near Mach 1, but the methodology used was then extended to airfoils for lower-speed application. The resulting airfoils were known as the *GA(W)* series, for "general aviation (Whitcomb)."

These airfoils were designed with computer technology by inputting the desired aerodynamic characteristics. Prior to such development, airfoil shapes were chosen first and then the aerodynamic characteristics determined from tests. The GA(W)-1 airfoil has already been incorporated in a number of designs, most notably Piper Aircraft Corporation's Tomahawk trainer (Fig. 2-33). It is identified by the marked concave shape of the aft lower surface. It is interesting to note that many modern light airplanes still use 6-series, five-digit, and even four-digit airfoils, despite the recent advances in this area.

Piper Aircraft Corporation

Fig. 2-33. Piper Tomahawk showing characteristic undercurvature of the GA(W) – 1 airfoil used in its wing.

## WING LIFT AND SPAN EFFECTS

Up to this point we have been discussing the airfoil, which is a two-dimensional section, or profile, of a wing. The profile shape has a great deal to do with the aerodynamic characteristics of a wing; however, we cannot have a real wing until we add a third dimension. That dimension is what we call *wingspan*. The length of the wing, or span, and the *planform* of the wing also affect the aerodynamic characteristics. Planform is the shape of the wing as viewed from directly above or below.

Remember that an airfoil creates a pressure differential between its upper and lower surfaces. This differential is analogous to a difference in electrical or magnetic charge. The result is a potential for inducing a flow from high pressure to negative (or lower) pressure. Along the span of the wing this tendency results in a pressure force being exerted against the wing surface as shown in Fig. 2-34A; however, at the tips of the wing, the pressure differential still exists, but no more wing exists to block the flow.

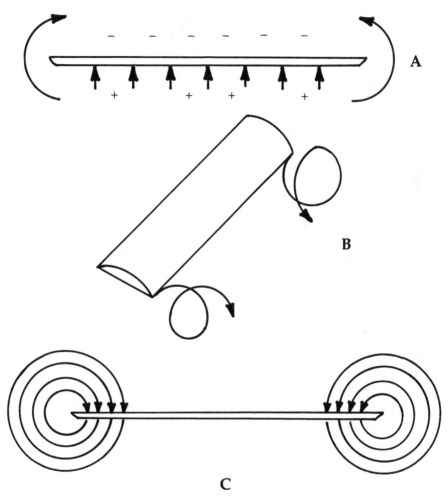

Fig. 2-34. Wingtip vortex formation induced by pressure differential (A) causing air to spiral about wingtips (B) with resulting flow shown in (C).

The result is a flow around the wingtip from the high-pressure area on the bottom surface to the lower-pressure area on the top. This motion is known as a *wingtip vortex*. As the wing moves forward this vortex trails on behind the wing and is also sometimes referred to as a *trailing vortex*. One emanates from each wingtip in a counterrotating fashion as shown in Fig. 2-34B, and the pair are usually spoken of in the plural as *wingtip vortices*.

Wingtip vortices are more commonly known to the pilot as *wake turbulence* and are usually thought of as something to avoid when shed by heavy aircraft.

The strength of the vortices is obviously proportional to aircraft weight because greater weight requires more lift and more lift requires a greater pressure differential across the wing surface. Remember, it is this pressure differential that creates the vortices, so when the differential is higher, the vortices will be stronger.

Vortices are not exclusively associated with heavy airplanes. Every aircraft wing generates vortices from its tips, although those produced by J-3 Cubs or Cessna 150s are not quite so dangerous and do not warrant strenuous maneuvering by pilots of following aircraft to avoid them. Furthermore, vortices have an effect not only on following aircraft, but also on the airplane that generates them. This is not a dangerous effect, but it does tend to make the wing less efficient than it would be if the vortices did not exist.

## Downwash effects

The tip vortices tend to exert a downward motion to the air leaving the trailing edge as shown in Fig. 2-34C. This downward push to the air is known as *downwash* and it has the result of changing the direction of the incoming airstream in the vicinity of the wing, as illustrated in Fig. 2-35. The downwash effect is greatest near the wingtips, but is experienced to some degree across the entire span. Because the lift is the force we create perpendicular to the airstream, or relative wind, we must consider the actual direction of the flow directly at the wing in order to determine the true direction of the lift force.

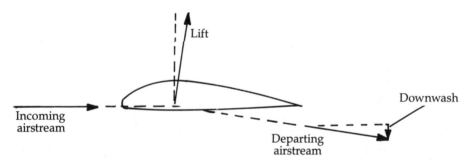

Fig. 2-35. Downwash pushing downward on airstream causing rearward tilted lift vector to result.

If the airstream is tilted downward by the downwash, the lift vector is then tilted somewhat aft. Part of the lift is now pulling in the streamwise direction and tending to retard the forward movement of the airplane. In so doing, some of the lift perpendicular to the incoming airstream is lost. If the airplane is flying straight and level, the weight will be downward perpendicular to the free airstream, so the lift to overcome it must be upward in a perpendicular direction.

When the lift vector is tilted backward, not all of the lift is acting perpendicular to the incoming airstream, as it would be without the downwash; therefore, a little more angle of attack is needed to make up for this loss of lift when downwash is present. This additional angle of attack is called the *induced angle of attack*, meaning that it is necessary because of the flow induced by the downwash.

Figure 2-36 shows a chart of the lift available for respective angles of attack if no downwash were present and another line for the lift *with* downwash. Notice that to get the same amount of lift with downwash, an additional angle of attack must be obtained: the induced angle of attack.

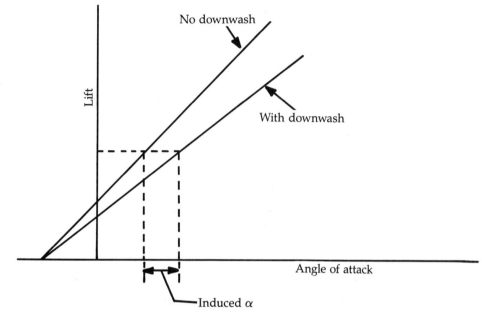

Fig. 2-36. Lift curves with and without downwash showing additional (induced) angle of attack required to generate equal amount of lift with downwash.

No downwash would only result if we had no tip vortices. This is actually a fictitious situation because it would require a wing of infinite span so that tips would not really exist. In reality, we have to have some finite span; hence, some downwash will always be present; however, the longer the span is, the more it will approach this ideal situation; thus, greater span reduces angle of attack required for a certain amount of lift. This reduces the drag associated with higher angles of attack and makes the wing more efficient.

## Aspect ratio

We could, of course, increase the span of a wing by simply changing only the span dimension, keeping the chord the same; however, this change would also increase the wing area. We could, on the other hand, keep the same area by increasing span and reducing chord proportionately. This would result in a skinnier wing. So that we can deal with wing area and wingspan separately, the term *aspect ratio* has been devised. Aspect ratio (AR) is defined as the span divided by the average chord. For a tapered or rounded planform wing, where the average chord might not be so apparent, the aspect ratio can also be determined by dividing the square of the span by the total wing area. The following formula gives two standard ways of determining AR.

$$AR = \frac{b}{c_{av}} = \frac{b^2}{S}$$

where   b = span
        c = chord
        S = wing area

Figure 2-37 shows two wings of different aspect ratio, but with the same area.

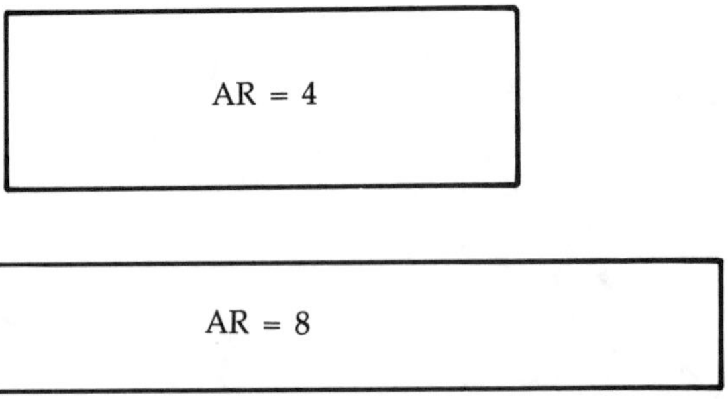

Fig. 2-37. Comparison of two wings with the same area but different aspect ratio.

## SPANWISE LIFT AND STALL SEQUENCE

If we had a wing of constant chord (rectangular planform), and if it could extend to infinity, we would have a constant amount of lift at each station along the span. Because infinite span is not possible, we have vortices and their associated downwash on every wing. The vortices tend to be strongest at the tips and have the effect of reducing lift most in this area. At the very tip of the wing, the pressures on the top and bottom surfaces have the potential to become equal, so it seems reasonable that the lift should go to zero; however, the vortex action causes a reduction in lift to extend inboard, and thus gradually reduces lift as spanwise direction increases toward the tip. Typical lift distribution across the span is shown in Fig. 2-38.

The lift at any spanwise station is proportional to the chord at that station. The lift is also dependent on the downwash at a particular station and this is affected by the planform shape of the wing. A rectangular wing, for example, has a greater concentration of trailing vortices near the tips than a tapered wing. The planform, or chord distribution along the span, therefore determines how the lift will be distributed along the span.

Because practically all wings are symmetrical about the airplane centerline, it is customary to display lift distributions on only one wing, or from the centerline to one tip. Such a wing portion is referred to as a *semispan*. Figure 2-39 shows the lift distributions over the semispans of a rectangular wing, an elliptical wing, and a tapered wing.

Because the lift is proportional to chord and angle of attack, the effective angle of attack of each spanwise section is also different. Figure 2-40 shows the

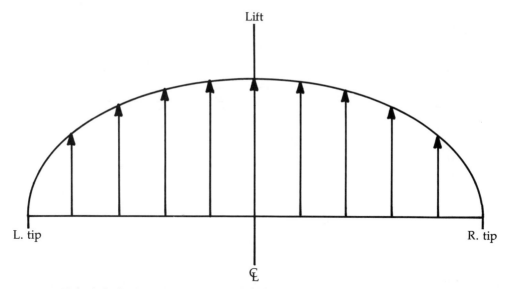

Fig. 2-38. Typical distribution of lift across the span of a wing.

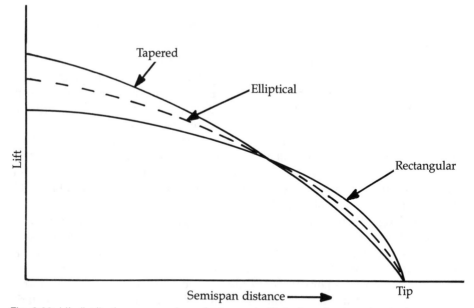

Fig. 2-39. Lift distribution on a semispan for tapered, elliptical, and rectangular planform shapes.

distribution of effective angle of attack of the three wing shapes considered. Notice that for the rectangular wing the section with the highest angle of attack is at the root. The tapered wing has the highest angle of attack at about two-thirds of the semispan. The elliptical wing has a constant angle of attack across the span.

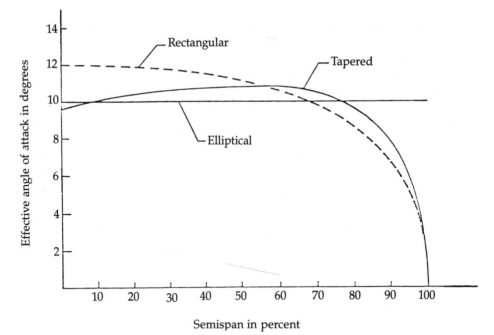

Fig. 2-40. Effective angle of attack across semispan for tapered, rectangular, and elliptical wings.

Because of the variation in angle of attack for different sections across the span, the stall will not occur simultaneously across the span. Remember that there is a definite angle of attack associated with stall for each airfoil section; therefore, assuming that all of the wing has the same airfoil, the station with the highest effective angle of attack will reach its stall point first. The elliptical wing should stall evenly across the span because its angle of attack is constant. The tapered wing should begin to stall on the outboard portion of the wing, and the rectangular wing should begin to stall at the root according to Fig. 2-40. The actual progression of flow separation for these three wings is shown in Fig. 2-41, and shows good agreement with these predictions.

The stalling sequence is of considerable significance. The tapered wing begins to stall first in its outboard region. This is very undesirable because loss of lift here means the potential for a roll to be induced. It is quite unlikely that the wing would stall exactly evenly on both sides of the centerline. Furthermore, the surface for roll control—namely, the *aileron*—is also located in this region, so that recovery from the induced roll would be impaired. This situation is particularly undesirable when close to the ground, which is exactly where we need to bring the wing up to (or close to) stall when landing. It would appear that landing a tapered-wing airplane could result in a day somewhat less than nice.

But now look at the rectangular wing. This sometimes esthetically berated form—the "Hershey-bar wing" of early Cherokees—appears to stall first at the root, right where we want it to. The outboard portion continues to fly right up to the break of the stall, giving us full aileron control throughout. Figure 2-42 shows

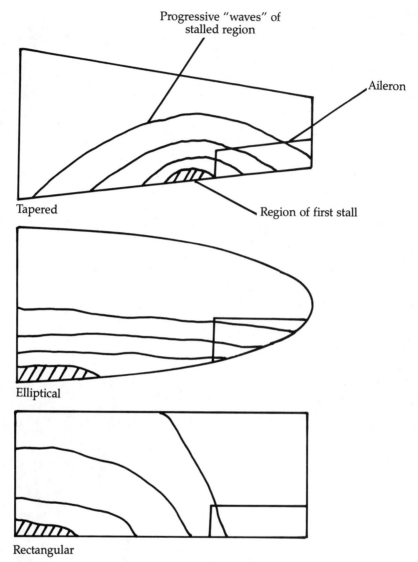

Fig. 2-41. Stall progression patterns for tapered, elliptical, and rectangular wings.

a Cherokee wing with yarn tufts attached to show regions of separated flow (stalled regions).

In normal flight, A, the tufts are all straight and aligned with the direction of flight. As the stall is approached, B, the inboard aft areas of the wing show separation as the tufts flutter around randomly; however, those in the outboard portion of the wing remain attached. Indeed, as I have witnessed numerous times, the tip area remains attached directly up through the deepest stall.

Many airplanes, however, do have tapered wings and they seem to be quite controllable laterally throughout the stall. The reason is that the wing has twist

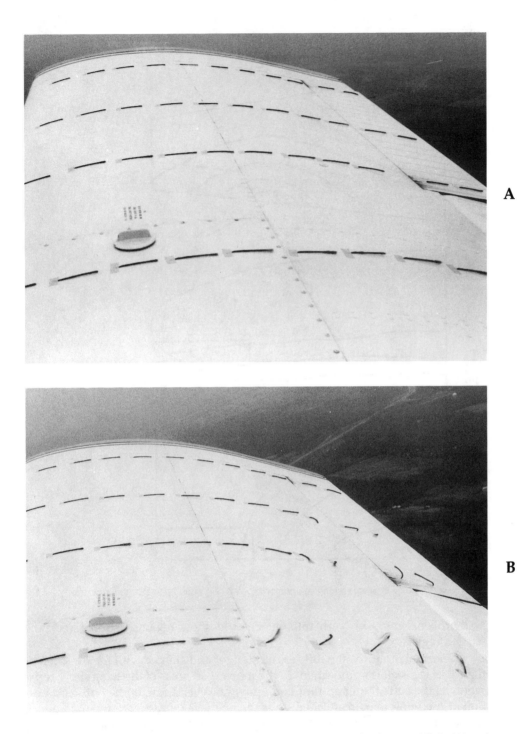

A

B

Fig. 2-42. Cherokee wing with yarn tufts attached to indicate flow condition in normal flight (A) and near stall condition (B).

built into it. The tip area is built so that it is set at a lower angle of incidence than the inboard sections. This is sometimes referred to as *washout*. Such configuration allows the inboard sections to reach stall angle before the outboard sections and forces the wing to stall in the desired sequence. One undesirable consequence of twisting is that it places some wing sections at angles of attack other than optimum, creating more drag.

Another way of attacking this problem is to use airfoils with higher stalling angles of attack in the outboard sections. This is referred to as *aerodynamic twist* because it achieves the same end as geometric twist. The inboard sections, with their lower stall angles, will begin to stall first.

## LIFT COEFFICIENT AND LIFT QUANTITY

Recall that the dynamic pressure possessed by a moving fluid was equal to $1/2\ \varrho$ $V^2$, or $1/2$ times the density times the square of velocity. So much pressure exerted over an area yields a force proportional to the amount of the area.

$$\text{force} = \text{pressure} \times \text{area}$$

The amount of lift obtained from a wing should, thus, be proportional to the dynamic pressure and the wing area. It is not exactly equal to the product of these two quantities, however. The portion of the resulting force being transformed into lift is measured by a *lift coefficient*. The lift coefficient is designated by the term $C_L$; thus, the general formula for lift is:

$$\text{lift} = C_L \times \left(\frac{1}{2}\varrho\ V^2\right) \times S$$

S, remember, was the symbol for wing area. The equation thus says that lift is equal to the lift coefficient times the dynamic pressure times the wing area. Lift coefficient can be thought of as a measure of how efficiently the wing is transforming dynamic pressure into lift.

Another way to think of lift coefficient is to consider a wing generating a certain amount of lift. Suppose, for example, that we tested a model wing in a wind tunnel. We could measure the lifting force on the wind tunnel balance. We could also measure the area of the wing and could determine the density and velocity of the air by appropriate instruments. The lift coefficient would be determined by dividing the measured force (lift) by the dynamic pressure and the wing area.

$$C_L = \frac{\text{lift}}{1/2\ \varrho\ V^2 \times S}$$

Assume that our wing has an area of 2 square feet, is subjected to a dynamic pressure of 1.5 pounds per square foot, and yields a lift force of 1.2 pounds. The lift coefficient would be determined as:

$$C_L = \frac{1.2}{1.5 \times 2} = 0.4$$

This result would be obtained, of course, at a certain angle of attack. At a higher angle of attack (but same wing and dynamic pressure) we would get a higher $C_L$. If we measured the wing lift at a number of angles of attack up through stall angle and calculated $C_L$s in this fashion, we could then get a complete curve of lift coefficient versus angle of attack. Figure 2-43 shows such a curve. The advantage of such a general lift curve is that for a certain angle of attack, it will tell you the lift coefficient available. You can then determine the lift available for *any* size wing at *any* speed by inserting the desired quantities into the general lift formula.

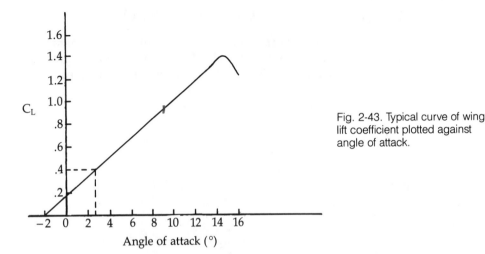

Fig. 2-43. Typical curve of wing lift coefficient plotted against angle of attack.

Lift coefficients for airfoils or wing sections can also be determined in this way. If a rectangular wing can be bounded by a wall (as it can be in a wind tunnel), it essentially has no tips and, therefore, simulates a section of an infinite wing. Measuring the lift in such manner and dividing it by the dynamic pressure and wing area then yields the lift coefficient associated with just the airfoil. There is no span effect because there is no tip to form a vortex and induce downwash. Airfoil data as developed by NASA, and NACA before them, was done in this way. The complete lift properties of the airfoil can be determined, then, by examining the appropriate plots of lift coefficient versus angle of attack, or *lift curve*, as it is known.

Figure 2-44 shows a typical airfoil lift curve or plot of $c_l$ versus $\alpha$. The small case c is used to refer to coefficients of airfoil sections, reserving the upper case C for wing properties. Several specific characteristics of the airfoil are of particular importance. One of the most important of these is the maximum lift coefficient ($c_{lmax}$). This is the lift coefficient at the stall point or peak of the curve. For this airfoil it appears to be 1.4. This would indicate that more lift could be obtained before stalling with this airfoil than one with, say, a $c_{lmax}$ of 1.2. Another consideration is the angle of attack associated with the stall or $c_{lmax}$ point. For this airfoil

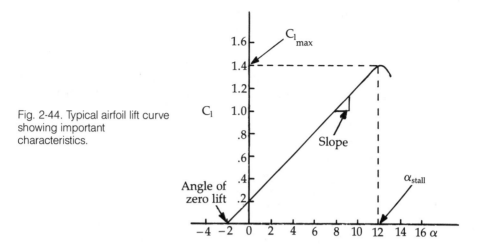

Fig. 2-44. Typical airfoil lift curve showing important characteristics.

it is 12°. A clue to the abruptness of the stall is also given by the sharpness of the peak of the lift curve. A very sharp dropoff indicates a sudden stall, while a rounded peak indicates a more gradual stalling of the airfoil.

This curve is for a cambered airfoil because some lift is developed at zero angle of attack. One must go to a negative 2° in order to develop zero lift. This angle where zero lift is developed is important also, and is referred to as the *angle of zero lift*. This angle would be zero for a symmetrical airfoil. The possible range of angles of attack available for flight is indicated by the difference between this angle and the stall angle. The slope that the lift curve possesses determines how rapidly $c_l$ increases with angle of attack. Below stall the curve is essentially a straight line, which indicates that $c_l$ is directly proportional to angle of attack.

All of these properties should be considered when selecting an airfoil for use in a design. NACA published the properties of numerous airfoils in a report, TR-824, in 1945. This information was later included in a book, *Theory of Wing Sections Including a Summary of Airfoil Data*, by Abbott and von Doenhoff, which is a very handy reference for aircraft designers (*see* bibliography). Examples of published airfoil data are included in appendix A.

# LIFT, FROM A
# MOMENTUM CHANGE CONSIDERATION

There is another way of explaining the creation of lift by an aircraft wing. This method involves the concept of *momentum*. Momentum is a physical quantity defined as mass times velocity. A mass of air moving at a certain speed has so much momentum determined by this product. Whenever momentum changes over a period of time, a force is exerted; force can be defined as the rate of momentum change.

Consider the incoming flow in Fig. 2-35. The air has momentum parallel to the flight path, or horizontal in the case of level flight; however, it has no vertical component; thus, the momentum in a vertical direction is zero because momentum must have magnitude and direction; however, when the flow leaves the wing, it has some downward motion due to the downwash, and now has some vertical momentum; therefore, the wing has changed the momentum of the air in a vertical direction and thereby created a vertical force downward on the air. According to Newton's Third Law, every action must have an equal and opposite reaction; thus, the air must exert an equal force upward on the wing. This, indeed, does occur.

Occasionally one might hear, among the hangar crowd, that this is the *true* explanation of lift and that Bernoulli's principle and the lift theory associated with it are just myths. This is not correct. The two methods of explaining lift are not opposing theories. They are merely different ways of looking at the same actions. Remember that the momentum change comes from the downwash, which is caused by the wingtip vortices. The vortices, in turn, exist because of the pressure differential between the upper and lower surfaces, and this differential is explained by Bernoulli's theory. The two approaches mutually support each other, and simply address the same physical phenomenon in different ways.

## FLAPS

In our original discussion of how lift is created by an airfoil, we saw that the cambered airfoil, even at zero angle of attack, creates a pressure differential between upper and lower surfaces. This pressure differential resulted, ultimately, from the greater area of the airfoil above the chordline interrupting more of the flow, as shown in Fig. 2-11. This greater upper-surface area results from camber. Obviously, the more camber there is to the airfoil, the greater this effect will be, and the more lift will be created. It follows, then, that higher camber in an airfoil creates greater lift.

Unfortunately, high camber also produces high drag. There is a limit to the degree of camber that will result in more increase in lift than in drag. Sometimes we can tolerate high drag along with high lift, such as during landing operations. At other times, we would like drag to be as low as possible. It would be nice if we could alter the camber of a wing in flight.

Some flexible wing structures have been built that would allow the actual bending of the entire wing along the chord to give a significant camber change. The first airplanes of the Wright era exemplify this concept. Wilbur and Orville took advantage of the "wing warping" ability to effect roll control. Later, experimental airplanes were designed with thick airfoil sections that would bend. Most modern aircraft structures, however, will not allow the entire airfoil to bend along the chord to a degree that will significantly affect the camber.

A very possible—and also practical—way of increasing camber is to hinge the airfoil at some point along the chord and bend just part of it. This approach allows for rigid structure and results in a surface known as a *flap*. The flap is the movable portion of the airfoil that is deflected through some angle from the original chord position to yield a higher camber.

Figure 2-45 shows an airfoil with a trailing-edge flap in normal (or retracted) position and in the deflected (or extended) position. In the deflected position, the camber is increased because the effective chordline is the line from the leading edge to the trailing edge as shown in Fig. 2-46. The camber is the deviation of the midline from the chordline, which can be seen to be quite large in this case.

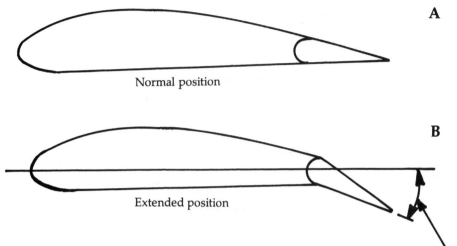

A

Normal position

B

Extended position

Angle of deflection

Fig. 2-45. Trailing edge flap in normal or retracted position (A) and extended position (B).

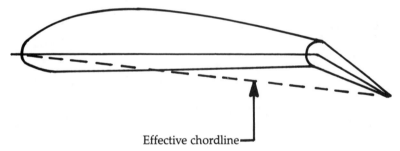

Effective chordline

Fig. 2-46. Effective chordline of wing with flap deflected.

The flap shown in Fig. 2-45 is called a *plain* flap. It is simply a part of the airfoil shape that is a separate structure and hinged so that it can be deflected. Other types of flaps are shown in Fig. 2-47. One in which only the lower surface of the airfoil deflects is called a *split* flap. Plain and split flaps are the simplest types of flaps from a construction standpoint. They are rather limited, however, in the additional amount of lift that they can create. At larger angles of deflection they create large wakes, thereby producing considerable drag and little additional lift.

An improvement on these basic designs is the *slotted* flap. Slotted flaps move

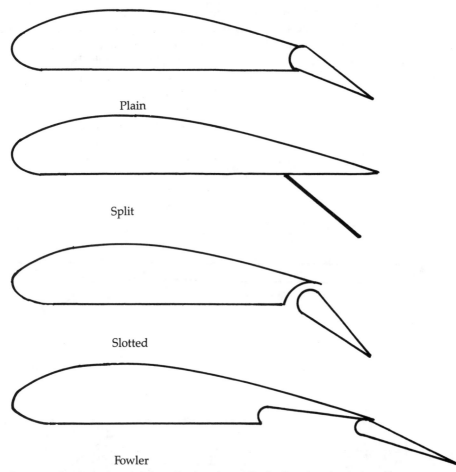

Plain

Split

Slotted

Fowler

Fig. 2-47. Configuration of plain, split, slotted, and Fowler flaps in extended position.

slightly aft as they deflect, opening a small slot from the lower to the upper surface. High-pressure air from the lower surface thus flows through the slot to the upper surface, providing more energy to the air flowing over the flap. This energy delays the flow separation and results in a smaller wake and less drag. The slotted flap can therefore be cranked down to fairly high angles, giving a good deal of additional lift without excessive drag.

Another rather special type of flap moves aft a considerable distance as it moves down, thereby increasing the wing area, called a *Fowler* flap. It requires a track or other mechanism on which to move and thus complicates the wing structure. Airplanes that require extreme increases in lift, such as jet airliners, sometimes employ combinations of various types of flaps. The slotted-Fowler flap is a high lift producer, and the double-slotted-Fowler combination is even more effective. The structure, however, is proportionately complicated.

Flaps increase the lifting capability of the airfoil with all other conditions being equal. What this means is that they increase the *lift coefficient*, the measure of lift efficiency. Figure 2-48 shows typical lift curves (plot of $C_1$ versus angle of attack) for an airfoil in the unflapped and the flap-deflected condition. Note that for any particular angle of attack (5°, for example), a greater lift coefficient is obtained with flaps down. This fact in itself is not too significant because a higher $C_1$ could be obtained without flaps by just going to a higher angle of attack. But higher angle of attack requires lower velocity because lift is proportional to angle of attack times the square of velocity.

$$L \propto \alpha \times V^2$$

To provide lift for a certain weight and airplane configuration, if one of these terms is increased, the other must be reduced to result in the same product. Lowering velocity would increase angle of attack (and corresponding lift coefficient), but would also approach—and eventually reach—the stalling point.

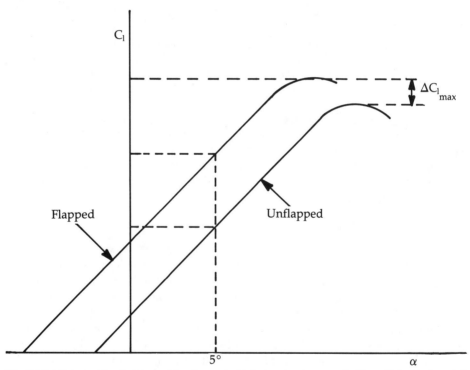

Fig. 2-48. Lift curves for flapped and unflapped airfoil showing increase in lift coefficient below stall and increase in maximum lift coefficient.

The really significant effect of flaps is to allow the lift coefficient to increase beyond its maximum in the unflapped condition. Maximum lift coefficient is reached at the stall angle. In the flapped case, even though stall occurs at a lower angle, the $C_1$ is much higher. The higher maximum lift coefficient allows the airplane to fly slower before it stalls (it effectively reduces its stall speed). Slower approach speed is therefore made possible, which allows slower touchdown speed, resulting in shorter landing distance.

Figure 2-49 shows the increase in lifting capability of a flapped wing in a wind tunnel. In Fig. 2-49A, the wing flap is fully retracted and the wing is set at near zero angle of attack. The lift is seen to be 0.26 pounds. In Fig. 2-49B, the wing is kept at the same angle of attack and the same airspeed, but the split flap is deflected 10°. Notice that the lift is increased to about 0.6 pounds. In Fig. 2-49C the flap is further deflected to 25° with all other conditions the same. The lift is now increased to 1.03 pounds. Note also, however, that the drag has just about doubled from the unflapped condition, while at 10° the drag increase was very slight.

These photos illustrate an important aspect of flaps. At small deflection angles they increase lift more than they increase drag. For this reason, small flap deflections can be helpful in shortening the takeoff run. Larger angles of flap deflection create so much drag that the takeoff roll is slowed to the point where the increase in lift is counteracted. In other words, the effect of the drag is greater than that of the lift.

In landing, however, increase in both lift and in drag is desirable. Increased lift lowers the stall speed (and, hence, landing speed) and increased drag steepens the descent angle, which aids in clearing obstacles near the end of the runway. Large flap angles are therefore favorable for this phase of flight.

## OTHER DEVICES FOR CONTROLLING LIFT

Most flaps are on the trailing edge of the wing. Flaps might also be on the leading edge of a wing, normally only used on large aircraft that require an extreme boost in lift coefficient for landing. Leading edge flaps are also usually used in conjunction with trailing edge flaps.

### Slots and slats

A more common device found on the leading edge is the *slot*. This device, as shown in Fig. 2-50, allows air to flow from the lower surface to the upper surface at high angles of attack. The higher pressure air from the lower surface has more energy, which will delay the separation of the airflow on the top surface and, thus, the onset of stall. It is another way of achieving higher lift at low airspeeds.

The disadvantage of the slot is that it creates excessive drag at lower angles of attack that are associated with normal cruise speeds. A way of avoiding this situation is to have a leading edge section that will open into a slot at low speed, but close at high speed. Such a movable device is referred to as a *slat*. The slat adds a bit of complexity, but has one convenient trait: When allowed to float freely, it will open and close automatically, if properly designed. In high-speed (cruise) condition, the high pressure on the leading edge will push the slat shut. At high angles of attack, when the stagnation point is low on the nose of the airfoil, the flow will be moving up over the leading edge, as shown in Fig. 2-50. The reduced pressure of this flow tends to pull the slat forward thereby opening the slot. Here is one situation where nature works with us to provide a desirable flight situation.

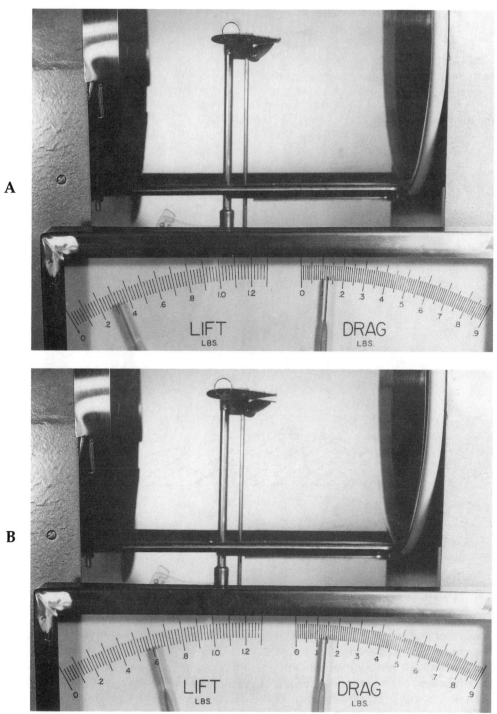

Fig. 2-49. Wing in wind tunnel with flap retracted (A), showing increased lift with slight flap deflection (B), and greatly increased lift with large flap deflection (C).

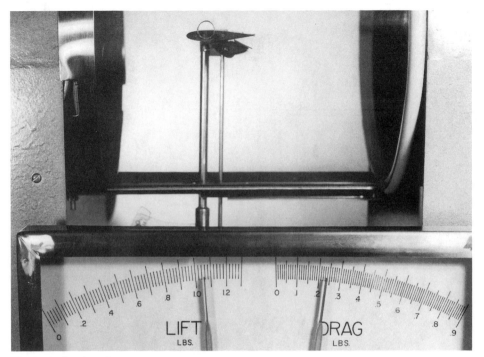

Fig. 2-49. Continued.

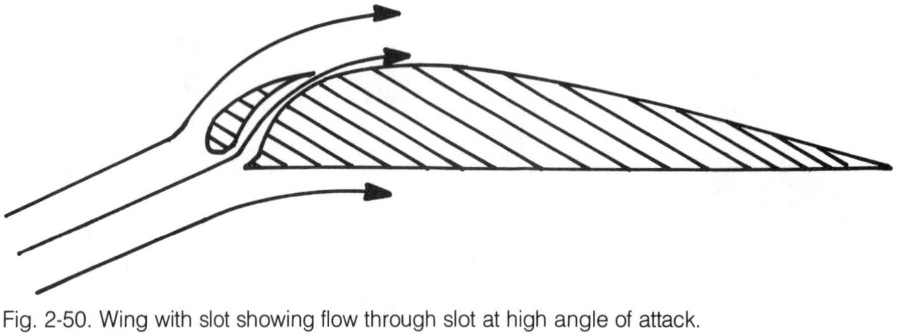

Fig. 2-50. Wing with slot showing flow through slot at high angle of attack.

## Spoilers

The *spoiler* is the opposite of a high-lift device; it is actually a destroyer, or spoiler, of lift, as the name implies. It is a device that affects lift and is in some uses indirectly involved in high lift.

Spoilers are very simple in their basic application. They project upward into the airstream, as shown in Fig. 2-51A, blocking the flow on the top surface, and thus destroying its lifting tendency. These devices are used on sailplanes to control sink rate. They are also used on large jet aircraft for the same purpose, in lieu of power reduction. There is considerable delay in regaining power on a jet after

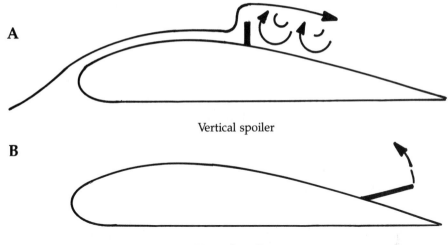

**A**

Vertical spoiler

**B**

Hinged spoiler

Fig. 2-51. A vertical spoiler projects upward into the airstream. A hinged spoiler acts somewhat like an aileron as it rises.

a reduction, so spoiler control is a much more rapidly responding method of affecting this action.

Another application of spoilers is their use as a primary roll control. Instead of deflecting ailerons, as explained in chapter 6, a roll can be induced by deflecting a spoiler on the outboard section of one wing. The lift is thus reduced on that wing, and the airplane will roll in that direction. The primary reason for using spoilers for roll control is that it frees the entire trailing edge span for flap use. Longer span flaps mean more flap effectiveness for lower speed capability.

A problem exists with this type of roll control in that the spoiler has to be extended some amount before it begins to take effect, and then the effect is rather rapid. In other words, there is a sloppy feel to the control stick at first, and then a very sensitive feel as roll is gradually increased. This effect can be overcome somewhat by hinging the spoiler to act somewhat like an aileron as it rises, as shown in Fig. 2-51B.

Using spoilers for roll control has another distinct disadvantage. That is that rolling can only be done by dropping one wing. If a wing drops close to the ground, it can be raised by increased lift on that wing, using normal aileron control. This effect cannot be created with spoilers, so low-level flight in gusty air must be done very carefully. Nevertheless, spoilers have been used rather successfully in this application. They will probably appear more frequently on new designs that attempt to expand the speed envelope at both ends, meaning higher maximum speed and lower stall speed.

# 3

# Drag

*DRAG* IS THE TERM USED TO DENOTE RESISTANCE TO AIRFLOW. It is the component of force created in the streamwise direction. In the process of creating lift, we also unavoidably create drag. Not only the wing, but all parts of the airplane create this unwanted opponent of flight.

Any physical body being propelled through the air will have drag associated with it. You merely need to hold your hand out of the window of a moving vehicle to experience this force. The actual physical processes that cause drag are a bit more complicated than this simple example. For this reason, drag deserves some discussion. There are also several distinct causes of drag and likewise several distinct names assigned to various types of drag, depending on their origin.

## PARASITE DRAG

Most of the drag experienced by holding your hand perpendicular to the ground outside a vehicle window to the airstream is *pressure drag*, so named because it results from the difference in pressure between the fore and aft sides of the hand. A very thin flat plate placed perpendicular in an airflow, as seen in Fig. 3-1A, would create purely pressure drag. The pressure against the upwind face is obviously greater than that in the wake formed behind the plate and a force in the streamwise direction results.

You might expect that if the plate were turned parallel to the airstream, as in Fig. 3-1B, that the drag would be eliminated, or at least be very, very slight. This is not the case. Actually, the air flowing along the surface of a body creates a frictional force on the body. This friction results from the fact that all fluids have some viscosity or stickiness. Air is usually not thought of as being sticky. It certainly is not as viscous as oil, or even water, but it does have some viscosity; therefore, it creates a frictional effect on a surface over which it flows. The flat plate parallel to the flow thus experiences *skin friction drag*, also termed *viscous drag*.

Most parts of an airplane are not extremely thin flat plates, however. They usually have thickness and surface area. Such bodies will exhibit pressure and

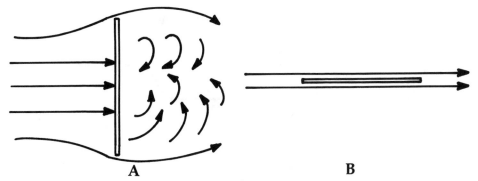

A                   B

Fig. 3-1 Flat plate aligned perpendicular to airstream (A), and parallel to airstream (B).

skin friction drag. The body in Fig. 3-2, for example, will have friction along its surfaces, but also some pressure drag because a wake is being formed behind it. These two types of drag are not entirely independent, either, because the size of the wake is dependent on the point of separation of the flow on the aft portion of the body. This separation point, as will be seen in further discussion, is, in turn, dependent on the viscous characteristics of the fluid; therefore, the combined drag resulting on such a body is given the term *parasite drag*. Parasite drag can be thought of, then, as the sum of pressure and skin friction drag.

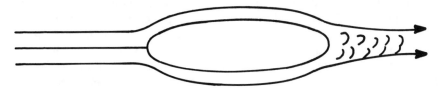

Fig. 3-2 Typical aerodynamic shape possessing both skin friction and pressure drag.

## SKIN FRICTION AND BOUNDARY LAYERS

Because the viscous effects of the air have a great deal to do with drag, they are worth discussing in some detail. The viscosity causes the air to stick at the surface of a body over which it flows; therefore, the velocity directly on the surface is zero for any velocity of the free airstream. Proceeding above the surface, the velocity gradually builds up until it reaches the free stream velocity at some distance above the surface. This distance is not very great for most flight conditions, but does have some definite thickness. This area between the surface and the point where the velocity reaches that of the free airstream is called the *boundary layer*.

Within the boundary layer the velocity takes on a profile similar to that shown in Fig. 3-3. It is gradually reduced from that of the free stream to zero right at the surface. The reaction to the retardation of the flow within the boundary layer is the skin friction drag. The boundary layer thus is the mechanism by which skin friction drag is created. The extent of the skin friction drag depends on the shape and the thickness of the boundary layer.

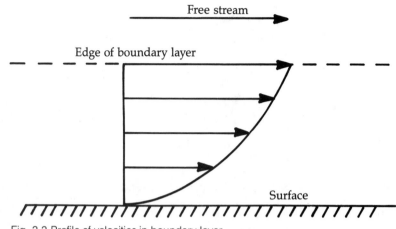

Fig. 3-3 Profile of velocities in boundary layer.

## Laminar and turbulent boundary layers

At the leading edge of a surface, the boundary layer, of course, has zero thickness. The thickness then increases as the flow progressively moves over the surface in the downstream direction. The thicker the boundary layer becomes, the more drag is created, because more momentum is being removed from the air. The amount of drag created also depends on the type of flow that exists. Fluids can flow in one of two ways. One is in a smooth, layered fashion, in which the streamlines all remain in the same relative position with respect to each other. This type of flow is referred to as *laminar flow*. The other type is *turbulent flow*, or one in which the streamlines break up and become all intermingled, moving in a random, irregular pattern.

The turbulent boundary layer is thicker than the laminar boundary layer (given the same relative free airstream), and thus creates more drag. It is obviously beneficial then to attempt to make as much of the air flowing over the aircraft as possible to flow in a laminar fashion. Such a goal might be noble, but not that easily achieved. Boundary layers seem to have a great desire of their own to go turbulent.

What starts out as a nice smooth laminar layer near the leading edge suddenly turns into a nasty, turbulent flow a short distance downstream. The area where a boundary layer changes from laminar to turbulent flow is called the *transition region*. Figure 3-4 shows a boundary layer starting out as laminar, going through transition, and then becoming turbulent.

This transformation in a flow can be seen in the smoke rising from a cigarette in calm air, as shown in Fig. 3-5. The smoke rises initially in a laminar manner. Then, as it encounters the friction of passing through the surrounding air, it transitions to a turbulent flow. This same effect takes place as a boundary layer moves over the surface of a wing or other aircraft component.

Any slight irregularity of the surface will tend to affect the boundary layer, particularly in the laminar region. Disturbances in this area will "trip" the

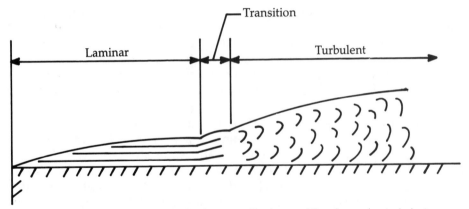

Fig. 3-4 Boundary layer beginning as laminar, transitioning, and then becoming turbulent.

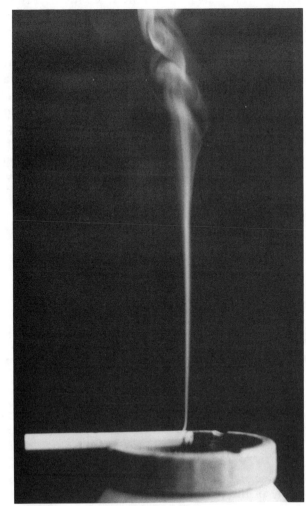

Fig. 3-5. Rising cigarette smoke beginning as laminar flow and then becoming turbulent.

boundary layer, as the disturbances are called, and cause the layer to prematurely transition to a turbulent one, thereby increasing drag. It is for this reason that much attention is paid to keeping leading edges of wings and other surfaces as clean as possible. Flush rivets are usually employed along with heavy coats of finish and fillers in crevices. Waxing can also be beneficial. Farther downstream these efforts have less effect, and that is why protruding-head rivets are often found closer to the trailing edges.

Even with the utmost attention to construction and finishing, however, attempting to retain laminar flow over an actual aircraft surface can be a very frustrating endeavor. Slight surface irregularities are always present, even on show aircraft in concourse condition. And laminar boundary layers can be even more finicky than Morris the cat. They display a great eagerness to turn turbulent after just a few inches of flow over a surface at normal flight speeds.

To give some idea of the extent and thickness of boundary layers, consider a 30-inch wide plate in an airflow at 135 mph. This is the chord length of the horizontal tail of a Piper Cherokee. Under normal conditions, the boundary layer would remain laminar to approximately 3 inches from the leading edge. At this point the layer would be approximately 0.03 inch thick. After going through transition to a turbulent layer, the thickness would jump to approximately $1/10$ inch and increase to slightly more than $1/2$ inch at the trailing edge.

A turbulent boundary layer has a different profile from that of a laminar boundary layer, however. As shown in Fig. 3-6, the reduction in velocity in the turbulent boundary layer is much less in the outer portion of the layer and much greater in the inner portion near to the surface. For this reason, the effective thickness of the turbulent boundary layer can be considered to be somewhat less than that of laminar layers. The really noticeable region of reduced velocity would be only approximately 0.07 inch, or less than $1/10$ inch at the trailing edge of the Cherokee stabilator for the airspeed stated.

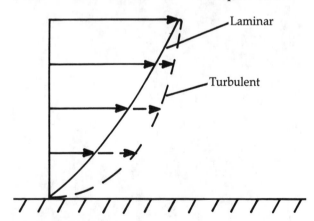

Fig. 3-6. Comparison of velocity profiles in laminar and turbulent flows.

## Reynolds number

A scientist by the name of Osborne Reynolds is responsible for discovering many of the principles of fluid viscosity and boundary layers. He found that whether the boundary layer was laminar or turbulent depended upon the fluid velocity, the distance downstream, and a characteristic of the fluid known as *kinematic viscosity*. A *Reynolds number* is used to measure the viscous qualities of a fluid. The symbol $R_e$ is used for this number and is defined as follows:

$$R_e = \frac{V \times d}{v}$$

where  V = Fluid velocity
            d = Distance downstream from leading edge
            v = Kinematic viscosity of the fluid

At low Reynolds numbers, the flow is laminar and at high Reynolds numbers it is turbulent; hence, from the formula, it can be seen that higher velocities or longer distances downstream tend to produce higher Reynolds numbers and consequently greater potential for turbulent flow. Because there is a slightly different Reynolds number at each downstream point, it is customary to use a *characteristic* length to define the Reynolds number for an airplane operating at a certain speed. The length normally used is the average wing chord. Based on this method of definition, an airplane with a wing of 5-foot chord going 100 mph at standard sea level would be operating at a Reynolds number of about 4.7 million.

## WAKES AND PRESSURE DRAG

We have already discussed briefly the phenomenon of pressure drag, which arises when a wake, or region of separated flow, exists behind a body. Now let us consider what causes the wake. In the case of the flat plate in Fig. 3-1A, it is rather easy to comprehend that the edges of the plate would cause the flow to separate at these points rather than flow around them. Nature seems to resist abrupt changes in trends and favors more gradual, smoother transitions. When a sharp change is introduced, nature reacts with something drastic. In the case of fluid flow, the drastic reaction is separation of the flow from the body and a wake, or region of relatively "dead" air, results.

It was pointed out before that the imbalance of pressure between the forward face of the plate and that on the aft face (in the wake) results in a drag force. Obviously, a progressively larger wake results in progressively more drag. If the plate were curved into the airstream, as shown in Fig. 3-7A, the wake would be larger than the flat plate of the same depth shown in Fig. 3-7B. Because the wake is larger, the drag on this plate would be greater. On the other hand, if the plate were curved in the other direction, as in Fig. 3-7C, the flow would be induced to form a smaller wake and the drag would be less.

Even curved downstream, however, the plate has a relatively large wake, and proportionate amount of drag. If this same plate had an afterbody to it that faired down gradually to a point, as in Fig. 3-7D, the wake could be reduced considerably. The trailing shape of a body has a great effect on the drag. In this case, it forms a mechanism to induce the flow along the trailing edge and gives it stream-

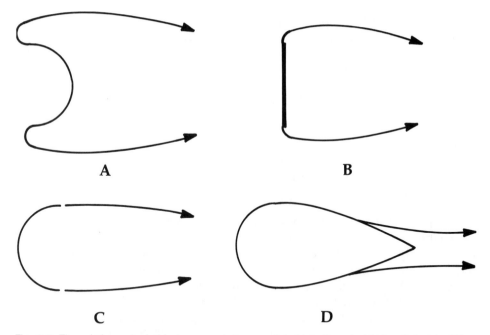

Fig. 3-7. The relative wake behind a curved shape pointed into the wind (A), a flat plate (B), a curved shape pointed downwind (C), and an aerodynamic shape (D).

lines in that direction. Unfortunately, the flow will not remain attached the whole way to the trailing edge, but will separate somewhat ahead of it, yielding a small but definite wake. This effect is further illustrated in Fig. 3-8. Replacing the circular flat plate (Fig. 3-8A) with the aerodynamic shape of the same diameter (Fig. 3-8B) reduces the drag by almost half.

The reason for this separation even where no sharp corner exists was mentioned in the discussion on stall, but can now be explained in greater depth. As the flow moves along the surface of the body it is creating friction from the boundary layer. As it progresses, the boundary layer is becoming thicker, and more friction drag is thus being created. Also, as the cross-sectional area of the body is getting smaller in the downstream direction, the velocity is becoming less, and consequently the pressure is increasing, previously explained as *adverse pressure gradient*.

The pressure is working against the flow rather than with it, as it is on the forward part of the body (ahead of the maximum thickness point). This combination of skin friction and adverse pressure gradient eventually gangs up on the airflow and prevents it from going any farther along the surface; it then has to separate. Exactly where this occurs depends on a number of qualities of the air and the body.

Usually we would like to delay separation as long as possible. The longer it remains attached, the smaller is the resulting wake and pressure drag; however, we must not forget that as long as the flow remains attached it is also generating skin friction drag. We want to limit these types of drag so that total parasite drag

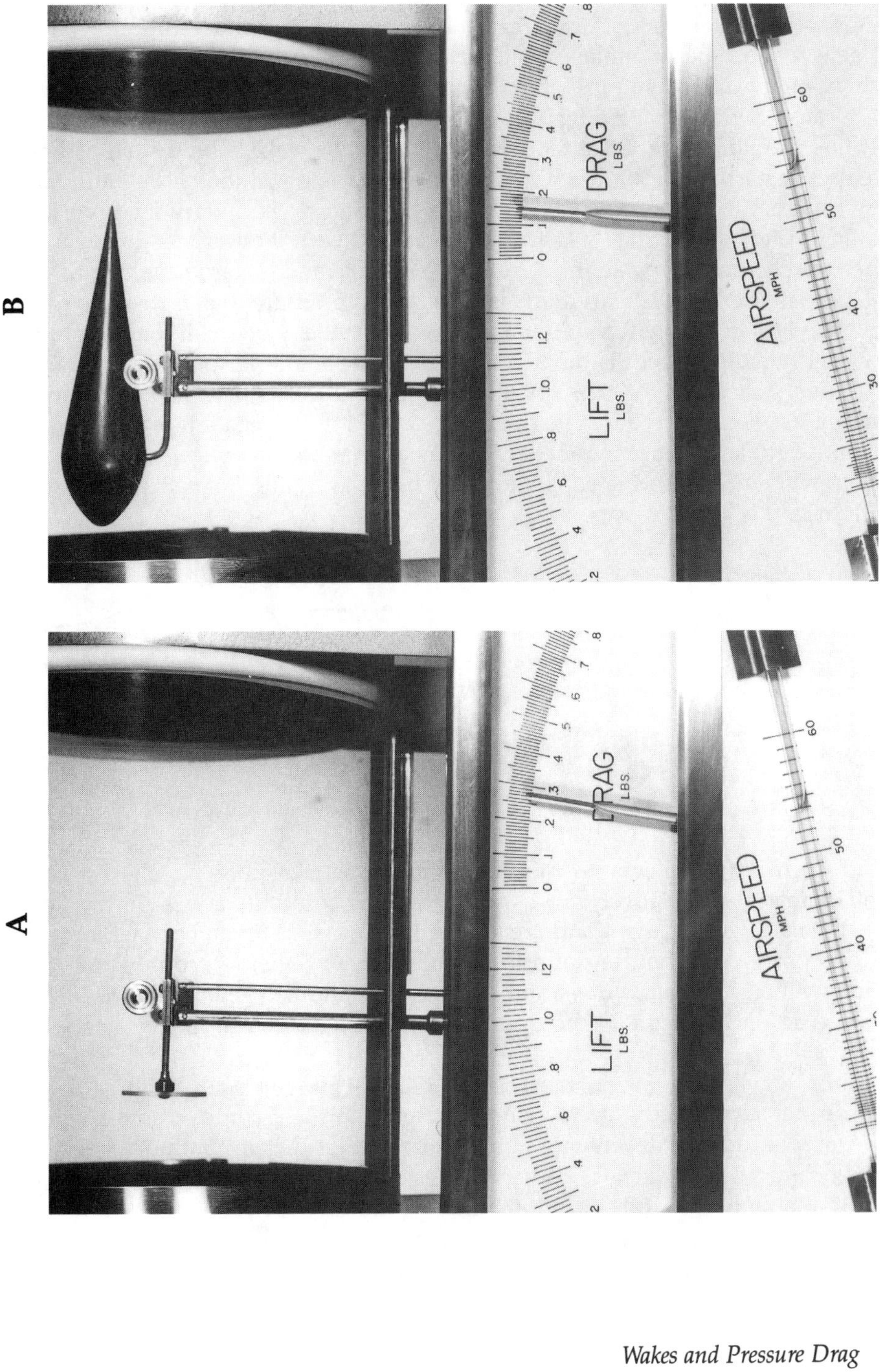

Fig. 3-8. Wind tunnel measurement of drag on a flat disc (A) and on an aerodynamic shape of the same maximum diameter (B).

is a minimum. It should also be obvious that these two types of drag are not independent because the skin friction induces separation and separation forms the wake that causes pressure drag.

For bluff bodies, or bodies with diameters that are fairly large in relation to their length, pressure drag is the greatest offender; fuselages, nacelles, and landing gear wheels are examples of such bodies. With thin, streamlined shapes, such as wings and tail surfaces, the drag is primarily skin friction drag.

Because bluff bodies suffer most from pressure drag, it would seem that attention should be directed at reducing the size of the wake behind such bodies. One trick that can be employed here is controlling the transition of the boundary layer from laminar to turbulent. Although the laminar boundary layer creates less *skin friction* drag, it has less energy associated with it and tends to separate rather readily. Early separation causes a larger wake and larger *pressure* drag.

The classic example of boundary layer effect on pressure drag is the sphere or circular cylinder. At low Reynolds numbers (low velocity or small diameter) the flow will remain entirely laminar and tend to separate early, forming a large wake, as shown in Fig. 3-9A. At higher Reynolds numbers, the flow will transition to a turbulent boundary layer before reaching the separation point. Turbulent boundary layers have more energy than laminar boundary layers; therefore, if the flow is turbulent, the separation will be delayed and a smaller wake will result, as in Fig. 3-9B.

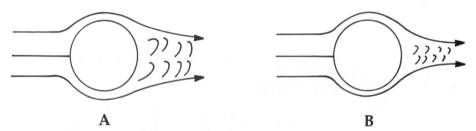

**A**　　　　　　　　　　　　　　**B**

Fig. 3-9. Relative wake behind a sphere with all laminar flow (A) and with partially turbulent flow (B).

If one had a low Reynolds number situation with a sphere, such as a very small diameter and relatively slow speed, the drag could be reduced if the flow were forced to go turbulent and create a smaller wake. Remember that surface condition can affect the transition to turbulent flow; therefore, roughening the surface will promote early transition and have the ultimate effect of lower drag. This is exactly the reason for the dimples on golf balls and the fuzzy surface on tennis balls.

Circular cylinders, incidentally, have a relatively large amount of drag, even with turbulent boundary layers. They are far from ideal aerodynamic shapes; however, production convenience has for many years dictated round wires and struts on many light aircraft. Early biplanes, with a maze of wires to provide structural support, were tremendous drag producers. As energy consumption has become a more and more important issue, you see many such shapes rapidly disappearing on even the slowest trainers. Fairing a circular cylinder into an aerodynamic shape can reduce the drag of the cylinder by as much as 700 percent at normal lightplane speeds.

# DRAG COEFFICIENT

Drag, like lift, is proportional to the dynamic pressure of the air and the area on which it acts. Also like lift, drag is not exactly equal to the dynamic pressure times the area, but varies with the shape of the body, surface roughness, and other factors; therefore, a drag *coefficient* is used to describe how much of the dynamic pressure force gets converted into drag. The equation is much like the lift equation, except that it measures the force in the streamwise direction, or parallel to the flow:

$$\text{Drag} = C_D \times \left( \frac{1}{2} \varrho \, V^2 \right) \times A$$

where   $C_D$= drag coefficient
   $\varrho$ = density
   $V$ = velocity
   $A$ = area

The term $1/2 \, \varrho \, V^2$, remember, makes up the dynamic pressure, and is sometimes collectively referred to by the single symbol q. Thus, using this notation:

$$\text{Drag} = C_D \times q \times A$$

The drag coefficient is quite analogous to the lift coefficient, which is a measure of how much of the dynamic pressure gets converted into lift. In a way, it is a measure of how efficient the wing is at creating lift. Similarly, the drag coefficient is a measure of how well the wing (or whatever body concerned) converts dynamic pressure force into drag. Unlike lift, however, we usually want drag to be as low as possible. *Low* drag coefficients are therefore the desirable condition, and efficiency is determined by how *little* of the pressure force is turned into drag.

The drag coefficient can also be thought of as the ratio of drag force to dynamic pressure force, or:

$$C_D = \frac{\text{Drag}}{q \times A}$$

It is this form of the equation that is often used to calculate $C_D$ from wind tunnel tests. The drag force is measured by some sort of scales or balance. This force is then divided by q × A, which is determined from a measurement of the airspeed, density, and area of the body.

One small problem arises in drag coefficient definition, and that is the area to be used. The drag is actually being generated by a three-dimensional body, yet the drag is proportional to only two dimensions of the body. For example, the fuselage shape shown in Fig. 3-10 has one projected area when viewed from the side, another when viewed from the top, and a third when viewed from the front. We could actually choose any one of these areas to represent the relative size of the fuselage in calculating a drag coefficient; however, we would have to be careful then to use that same view to determine the area of any other fuselage if we wanted to calculate its drag by using our drag coefficient determined in this manner.

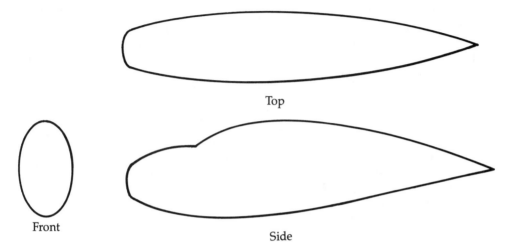

Fig. 3-10. Projected area of front, top, and side views of a fuselage.

The customary procedure for bluff bodies (such as the fuselage, nacelle, or landing gear), is to use the projected *frontal* area, or maximum cross section. For thin-profile bodies, such as the wing or tail surfaces, the planform (top view) area is normally used. Another area that is sometimes used is known as the *wetted area*, which is the total surface area of the body. This is the area that would be wetted if it were immersed in water, hence, the reason for the term.

Sometimes, in the case of components of complicated shape, a somewhat fictitious area is used. For example, the drag coefficient of an entire landing gear leg, including wheel, strut, and fairings, can be based on an area equal to the width times the diameter of the tire. This area is, essentially, the projected frontal area of the tire, but not quite. In early automobile aerodynamics it was customary to base the drag coefficient on a rectangular area equal to the overall width of the vehicle times the height.

Again, this area is not quite the projected frontal area, but is representative of the size presented to the airstream, and is also easy to calculate. Figure 3-11 shows the comparison of such areas for a typical car. Obviously, the drag coefficient of a Corvette or Porsche determined in this manner would be lower than that of a van, but it would also be lower for the more aerodynamically efficient car, even when compared against a van of exactly the same width and height. The reason is that the actual pounds of drag force would be less. Remember that drag coefficient is the actual drag divided by dynamic pressure and representative area:

$$C_D = \frac{D}{q \times A}$$

Therefore, shapes of the same area in a stream with the same dynamic pressure will have different drag coefficients because they will generate different amounts of drag. The more efficient shapes will have lower drag and hence lower drag coefficients. Incidentally, auto makers now base $C_D$s on actual area.

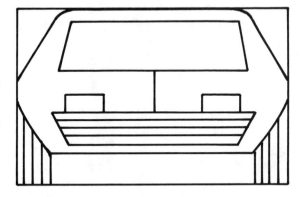

Fig. 3-11. Rectangular area made up of a maximum width times maximum height, once used to measure the drag coefficient of ground vehicles.

To illustrate this process of just how drag coefficients can be determined, let us use a hollow hemisphere 3 inches in diameter. We will measure the drag of the hemisphere in a wind tunnel and determine its drag coefficient. First the hemisphere is tested with its open, cup-shaped side to the airstream, as shown in Fig. 3-12A. The wind is moving from left to right in the wind tunnel pictured here.

Notice that the velocity meter indicates a wind of 53 mph (77.75 ft/sec), which translates to a dynamic pressure (q) of 6.85 pounds per square foot when combined with the air density of 0.002265 slugs per cubic foot. The drag is indicated as 0.52 pounds and the 3-inch diameter hemisphere has a circular cross-sectional area of 7.07 square inches or 0.049 square feet. The drag coefficient can thus be determined as:

$$C_D = \frac{D}{q \times A} = \frac{0.52}{6.85 \times 0.049} = 1.55$$

Now we turn the hemisphere around in Fig. 3-12B so that its rounded side is pointed into the wind. The same measurements are recorded. The air velocity is still 53 mph, thus yielding the same dynamic pressure, but the drag is seen to be only 0.24 pounds. The drag coefficient in this case is calculated as:

$$C_D = \frac{D}{q \times A} = \frac{0.24}{6.85 \times 0.049} = 0.715$$

We have reduced the drag coefficient of the hemisphere by more than half, or more than doubled its efficiency, by simply reversing its orientation. It was the same hemisphere, so that its area was the same, and the air velocity remained

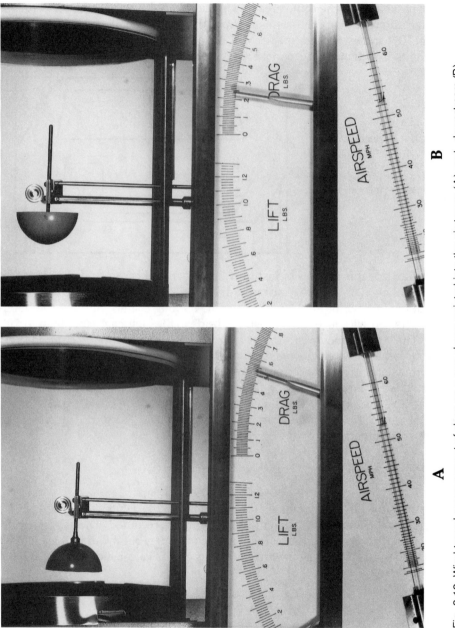

A

B

Fig. 3-12. Wind tunnel measurement of drag on a cup shape pointed into the airstream (A) and downstream (B).

unchanged. We have thus shown how drag is affected by variation in shape and how the relative drag efficiency is reflected in the drag coefficient.

The great usefulness of the drag coefficient, however, is that once determined, it can then be used to calculate what the drag would be with other areas or velocities involved. For example, suppose we wanted to determine the drag of a hemisphere 10 times the area of the one tested and at 100 mph (or 146.7 ft/sec). Using the latter, or most efficient orientation, where the $C_D$ was 0.715, the drag for this new situation could be calculated by inserting the appropriate terms into the general drag equation:

$$D = C_D \times \left( \frac{1}{2} \varrho\, V^2 \right) \times A$$

Remember that terms must be in consistent units, so that V must be in feet per second, $\varrho$ in slugs per cubic foot, and A in square feet. If you work this out on your calculator using the same density as in the test (0.002265 slugs/ft$^3$) you can predict the drag to be approximately 8.54 pounds.

# INDUCED DRAG

Chapter 2 discussed the mechanism for producing downwash. Figure 2-34 depicts the pressure differential between the upper and lower surfaces of the wing that induces a vortex to form at each tip, causing a downward push on the air leaving the trailing edge. This downward component, known as downwash, causes the airstream to depart at an angle downward from the incoming air as shown in Fig. 3-13. The lift vector, being perpendicular to the actual airflow, is thus tilted backward, resulting in a component of lift in the streamwise direction. A force in this direction, or opposite to the path of flight, is by definition drag. This particular drag, which results directly from the production of lift, is known as *induced drag*.

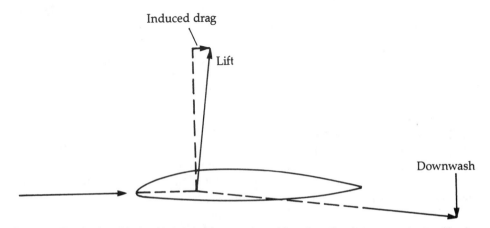

Fig. 3-13. Production of induced drag by downwash pushing down the airstream vector resulting in tilted lift vector.

Induced drag plays a large role in the performance of an airplane. It is, unfortunately, greatly misunderstood by many aviators, and often glossed over in pilot training manuals simply as "the drag due to lift." Induced drag is always present to some degree, although its effects can be altered somewhat. Because induced drag results directly from the wingtip vortices, it will be around as long as we have wingtips. The only way to completely eliminate it would be to have a wing of infinite span, which is obviously not a very practical design.

Longer wingspan (although never even near infinity) does place the wingtips farther apart and has the effect of reducing downwash and the resulting induced drag. The wingspan-to-chord ratio, known as aspect ratio (discussed in chapter 2), is a more useful way of describing span effects; thus, higher aspect ratio wings have less induced drag associated with them.

Induced drag can also be described or measured by a drag coefficient in the same manner as parasite drag. In the case of induced drag, the coefficient is, of course, referred to as the *induced drag coefficient*, and is the ratio of induced drag to dynamic pressure times area. It is noted as $C_{D_i}$.

$$C_{D_i} = \frac{D_i}{q \times A}$$

Induced drag, $D_i$, can then be determined by rearranging this equation to the form:

$$D_i = C_{D_i} \times q \times A$$

Because induced drag results from lift production, which is proportional to wing area, the area normally used for induced drag determination is the wing area, S.

Also because the induced drag is caused by lift, it could be reasoned that the induced drag should be proportional in some way to the lift coefficient. Actually, it turns out that it is proportional to the square of the lift coefficient. It is also inversely proportional to the aspect ratio, as was reasoned before. The induced drag coefficient can therefore be defined:

$$C_{D_i} = k \times \frac{C_L^2}{AR}$$

The k is a constant of proportionality and varies slightly with the planform shape of the wing and its orientation with respect to the fuselage. For an elliptical-shaped wing with no fuselage present, k is $1/\pi$ (where $\pi$ is the mathematical ratio equal to approximately 3.1416). This is the minimum value that k can have, and therefore implies that an elliptical wing has the lowest induced drag. Any other shape will have a larger k, and thus larger induced drag.

This fact was known pretty far back in aviation history, and for many years, much effort was devoted to designing wings of elliptical or near-elliptical shape. The British Spitfire of WWII fame (Fig. 3-14) and its racing predecessors were

Smithsonian Institution

Fig. 3-14. British Supermarine Spitfire of World War II showing elliptically shaped wing.

notable examples of such design. Unfortunately, as labor costs soared, the cost of producing elliptical wings began to outweigh their advantage, and they have pretty much disappeared in modern designs. Tapered wings are almost as efficient but much easier to build, which accounts for their popularity.

The proportionality of induced drag coefficient to the square of lift coefficient has a very significant effect on induced drag. Lift coefficient, remember, was defined as the ratio of lift to q × S.

$$C_L = \frac{L}{q \times S} = \frac{L}{\left(\frac{1}{2}\varrho V^2\right) \times S}$$

Because the induced drag coefficient is equal to this whole term squared and multiplied by k/AR:

$$D_i = C_{D_i} \times \left(\frac{1}{2}\varrho V^2\right) \times S$$

$$= \frac{k}{AR} \left[ \frac{L}{\left(\frac{1}{2}\varrho V^2\right) \times S} \right]^2 \times \left(\frac{1}{2}\varrho V^2\right) \times S$$

The resulting equation ends up with the square of the velocity in the denominator, or precisely:

$$D_i = \frac{k}{AR} \times \frac{L^2}{\frac{1}{2}\varrho V^2 \times S}$$

This means that increased velocity reduces induced drag and reduced velocity increases it, because induced drag is *inversely* proportional to $V^2$.

We can rewrite this equation in terms of span, b, remembering that $AR = b^2/S$. Also, considering the fact that in steady, level flight the lift must equal the weight, we can replace lift with weight, W. The equation for induced drag then becomes:

$$D_i = \frac{k\,W^2}{\frac{1}{2}\varrho\,V^2\,b^2}$$

Induced drag is proportional to everything in the numerator, namely our factor k and the square of weight. It is inversely proportional to everything in the denominator, which includes the square of the velocity, the square of the span, and the density. High density occurs at low altitudes, which means that this region favors low induced drag. Higher altitude, on the other hand, has lower density, and thus *higher* induced drag.

Factors that tend to *increase* induced drag become apparent:

- High weight
- Less efficient wing design (than elliptical)
- High altitude
- Low velocity
- Low wingspan

Of course, those factors that occur as the square of the term—namely W, b, and V—will have a greater effect. Airplanes that fly at very high altitudes encounter relatively high induced drag; however, increased span will counteract the effect and this is the reason for high aspect ratios on high-altitude aircraft. The U-2 reconnaissance aircraft, for example, has an aspect ratio of 14.

## TOTAL DRAG

The previous discussion on parasite drag considered just a simple body shape such as a sphere or a streamlined pod. Such shapes do not generate any appreciable lift and thus the parasite drag is the total drag. Wings or other lifting surfaces, as we just saw, generate induced drag, and also have a certain amount of parasite drag. A total airplane, of course, has parasite drag, but must also produce induced drag when it produces lift.

It is customary when calculating airplane drag to consider these two types of drag separately; however, total drag is made up of the sum of the parasite and the induced drag. For a complete airplane, parasite drag is considered as being composed of all of the drag except the induced. As we shall see, there are some special types of drag included in this term for the whole airplane over and above simple pressure drag and skin friction.

We have shown that induced drag behaves contrary to common sense, and decreases as velocity increases. Parasite drag, on the other hand, behaves quite respectably, and increases with increasing velocity. If we need to reduce parasite drag, we simply slow down, and it drops off just as you might expect it to do. It also varies with the square of velocity, but directly so. The equation for total drag, therefore, is usually broken down into two terms, one for parasite drag and one for induced drag. Such an equation could be written as:

$$D_{total} = (C_{D_P} \times q \times S) + (C_{D_i} \times q \times S)$$

The first term is the parasite drag term and the coefficient is written as $C_{DP}$. The second term is the induced drag, using $C_{Di}$ for the parasite drag coefficient, as previously discussed. The wing area, S, is normally used as the representative area for parasite and induced drag for a complete airplane.

The total drag could also be written with the induced drag in the form lastly derived in the previous section. The drag equation would then be as follows:

$$D = C_{D_P} \times \left( \frac{1}{2}\varrho\, V^2 \right) \times S + k\, \frac{W^2}{\frac{1}{2}\varrho\, V^2 \times b^2}$$

If we kept all of the terms the same except the velocity (meaning constant weight, wing span, area, and density), we could group all of these terms into two constants, $K_1$ and $K_2$. The equation now simplifies to the following:

$$D = K_1 \times V^2 + K_2 \times \frac{1}{V^2}$$

The first term is still the parasite drag term and can now very readily be seen to vary directly with the square of velocity. The second term is the induced drag and varies inversely with the square of velocity.

The drag variation with velocity could be plotted as shown in Fig. 3-15. The parasite drag starts off at zero at zero velocity and increases with $V^2$. Induced drag is infinitely high (hypothetically) at zero velocity and drops off inversely proportional to $V^2$. The total drag, however, would be the sum of the two curves and is shown by the dotted line.

Notice that there is a velocity point where the drag is minimum. This point occurs where the two curves cross, or where induced and parasite drag are equal. *Above* and *below* this point the drag increases. This particular fact has a great deal to do with the performance of an airplane. Range, rate of climb, and fuel consumption can be improved by slowing down the airplane—to a certain point. Below that point, performance again deteriorates. This, of course, is due to the fact that induced drag is building up more rapidly than parasite drag is dropping off. Chapter 5 continues this discussion with a detailed examination of these effects related to aircraft performance.

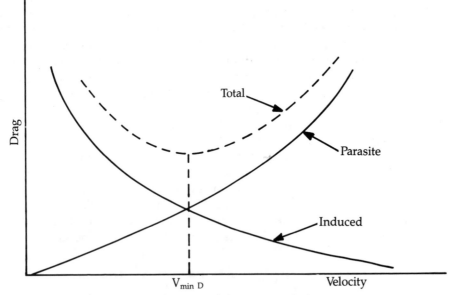

Fig. 3-15. Plot of induced, parasite, and total drag versus velocity.

# EQUIVALENT FLAT PLATE AREA

The parasite drag coefficient is a measure of the aerodynamic efficiency or aerodynamic "cleanness," as it is sometimes called, of an airplane. Comparing the $C_{D_p}$ of one airplane against those of others of the same general size gives an indication of its relative performance potential.

Another such indicator of drag efficiency is what is known as *equivalent flat plate area*. In utilizing this method of measuring drag, the parasite drag of the airplane is represented by the drag of a flat plate with a specified area that is oriented perpendicular to the airstream. The plate is also assumed to have a $C_{Dp}$ equal to 1. This is actually a fictitious situation because in reality a flat plate held against the airstream has a drag coefficient of 1.18. For this reason, flat plate area is more properly termed "equivalent *parasite* area." The symbol of f for this area, however, came from the original adjective, and the term "equivalent flat plate area," or sometimes, just "flat plate area" is still used by some of our most distinguished aeronautical scientists.

Equivalent flat plate area is really the ratio of parasite drag to dynamic pressure.

$$f = \text{flat plate area} = \frac{D_p}{q}$$

Remember that the parasite drag coefficient was defined as the ratio of parasite drag to dynamic pressure *times* the area (wing area, in the case of a complete airplane).

$$C_{D_P} = \frac{D_p}{q \times S}$$

Thus, the flat plate area is this equation rearranged slightly with the area on the left side. This arrangement shows that flat plate area is then the product of parasite drag coefficient times wing area.

$$f = C_{D_P} \times S = \frac{D_p}{q}$$

The parasite drag coefficient and the flat plate area are related through the wing area. If $C_{Dp}$ is known, f could be determined by multiplying by wing area; if f were known, $C_{Dp}$ could be determined by dividing by wing area.

To better understand the meaning of flat plate area, look at Table 3-1. This table lists the flat plate areas and also the parasite drag coefficients of 10 light airplanes. These values do not represent official manufacturers' engineering data, but were calculated from performance data published by the respective companies. As such, they are only approximations.

## Table 3-1. Parasite drag coefficients and flat plate areas of 10 lightplanes.

| Aircraft | $C_{D_P}$ | Flat plate area (sq. ft). |
|---|---|---|
| Cessna 152 | 0.038 | 6.14 |
| Piper Tomahawk | 0.054 | 6.64 |
| Beech Skipper | 0.049 | 6.36 |
| Cessna 172 | 0.036 | 6.25 |
| Piper Warrior | 0.034 | 5.83 |
| Beech Sierra | 0.034 | 5.02 |
| Mooney 201 | 0.017 | 2.81 |
| Piper Arrow | 0.027 | 4.64 |
| Cessna 182 | 0.031 | 5.27 |
| Beech Bonanza | 0.019 | 3.47 |

Consider the case of the Cessna 152, which shows an f of 6.14. This means that the parasite drag of the 152 could be duplicated by a flat plate of 6.14 square feet, providing that the plate had a drag coefficient of exactly 1. It does *not* mean that the total cross-sectional area of the Cessna 152 is 6.14 square feet.

Notice, for example, that the Skylane, which is a larger airplane, actually has a *smaller* f (5.27). The Skylane's parasite drag could be represented by this smaller flat plate area. Because it is a larger airplane, it must be aerodynamically cleaner in order to have this relative amount of drag. Notice also that the Bonanza and the Mooney have even smaller f values. These are retractable-gear airplanes, and both are known for their low-drag—and corresponding high-performance—characteristics. The Mooney (Fig. 3-16) is also pretty small for a four-place air-

Fig. 3-16. Low drag configuration of the Mooney 201.

plane, which, combined with its very clean design, gives it about the lowest parasite drag of current standard production airplanes.

It has been reasoned by those in the high places of aerodynamic circles that an f of 2.00 square feet is about the limit that can be obtained for a two-place or larger airplane, using current state-of-the-art production methods. Some homebuilders and disciples of certain sport airplane designers will dispute that claim; however, it applies pretty well to aircraft of conventional design. Sport and homebuilt airplanes are specialty items and indeed can surpass this drag barrier with very small wing areas, prone-positioned occupants, and other drag-reducing design schemes.

The $C_{Dp}$ values are also listed in this table. You can see that these figures also show a very good comparison of the relative aerodynamic efficiency of various designs. Remember that these values are determined simply by dividing the flat plate area by the wing area.

One other advantage of using flat plate area to calculate drag is that it is a convenient way to sum the drag of various components. Drag coefficients for the same shape can vary, depending on which area is used as the representative area; however, flat plate area is the product of drag coefficient times the respective area, so it should be the same for the same component, regardless of the area used.

To illustrate, consider the pod-shaped body in Fig. 3-17. Assume that we have measured the drag on this body at 2 pounds in a dynamic pressure of 4 pounds per square foot. If we used the planform area of 2.5 square feet (Fig. 3-14A) as a representative area, the drag coefficient would be as follows:

$$\text{(a)} \quad C_D = \frac{D}{q \times A_a} = \frac{2}{4 \times 2.5} = 0.2$$

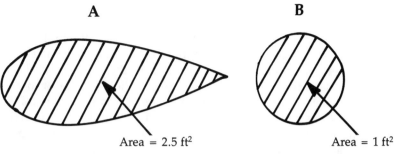

A                          B

Area = 2.5 ft$^2$            Area = 1 ft$^2$

Fig. 3-17. Projected area of top view (A) and front view (B) of a pod-shaped body.

We could also get a different drag coefficient if we used the frontal area as shown in Fig. 3-16B.

$$(b) \quad C_D = \frac{D}{q \times A_a} = \frac{2}{4 \times 1} = 0.5$$

Both drag coefficients correctly represent the drag characteristics of this body and could be used to calculate drag at other conditions. The proper area must be used, though, with the proper drag coefficient; however, if we considered the flat plate area, which is $C_D$ times A, it comes out the same.

$$f = C_{D_a} \times A_a = 0.2 \times 2.5 = 0.5$$
$$f = C_{D_b} \times A_b = 0.5 \times 1 = 0.5$$

Flat plate area can be used to add up drag of various components of the airplane. Wing drag could be calculated on the basis of wing area, and fuselage drag on the basis of frontal area. The resulting flat plate areas would not depend on the areas used, however, and could be added directly. Such a method of calculating drag for various components and summing them for total airplane parasite drag is known as a *drag breakdown*. The resulting value of f could then be divided by the wing area to yield a $C_{Dp}$ for the total airplane, which could be used in the parasite term of the drag equation to calculate drag at various airspeeds.

## SPECIAL TYPES OF PARASITE DRAG

All drag can be classified as either parasite or induced drag. Parasite drag can further be broken down into skin friction and pressure drag. It can be caused only by the forces arising either from the frictional effect of a viscous fluid or the pressure differential existing between the fore and aft surfaces of the body.

Parasite drag is created in a rather complex fashion in certain situations and the frictional and pressure effects are not very clearly explained. The parasite drag in such cases is often given a special term.

# Interference drag

One such type of drag is *interference drag*, which arises due to the juncture of two different bodies. Suppose you were to measure the drag on a wing in a flow of a certain dynamic pressure and then a fuselage in the same flow. If you then joined the wing to the fuselage and again measured the drag, you would find the drag to be somewhat greater than the sum of that of the individual parts. The reason is a formation of small vortices and interaction of boundary layers at the juncture. This effect can be reduced by filleting of the intersection, but never completely eliminated.

It is more critical, in the case of wings, on the upper surface, due to higher velocity flow in this region. It is therefore more important to fair in the upper surface of a wing on a low-wing airplane. In high-wing configurations, the upper surface is relatively undisturbed, and requires no filleting. The lower surface generates less interference drag at a junction, and thus is often neglected on high-wing aircraft.

Interference drag causes error in a drag breakdown calculation. Corrections are added to account for this. The amount of correction varies with the wing configuration, the degree of fairing, the number of junctures, and other factors.

# Cooling drag

Another specially identified type is *cooling drag*, which is associated with engine cooling. Practically all light airplanes are air-cooled and require a continuous flow of air over the cylinders and other engine parts to dissipate the excess heat. As the air flows over these parts, it encounters considerable drag from pressure and frictional effects. The air does gain heat because that is the purpose of directing it over the engine.

Heat is energy and gives potential momentum to the airstream. This is the principle on which turbojet engines are partially based and fuel is burned specifically to heat up the exhaust gas. Unfortunately, more momentum is lost in flowing through the engine compartment than can be added by heat. This difference results in a drag force.

Cooling drag has traditionally not been considered a large factor in holding back an airplane's forward progress; however, reductions in the parasite drag of nearly all other parts of the airplane have been brought about by great efforts in recent years to counteract high energy costs. As a result, cooling drag, which has not had a lot of attention until very recently, is now a significant factor. It is very difficult to calculate accurately, and even more difficult to reduce appreciably.

*Cowl flaps*, which adjust the flow of air over the engine to eliminate unnecessarily high heat under certain flight conditions, have been one way of reducing cooling drag. They can be opened wider for greater air volume flow during climb and closed down at higher-flow-rate conditions such as cruise and descent. Their presence, of course, adds increased complexity and consequent cost—and also some weight—to the airplane. Undoubtedly, with research efforts directed in this area, some new designs for natural cooling drag will emerge in the coming years.

# Profile drag

Another drag term that should be mentioned is *profile drag*, which is simply the wing's parasite drag. With so much concern about wing design effects on induced drag, we should not overlook the fact that the wing also generates parasite drag. Much of the wing's parasite drag is skin friction drag because of its large surface area; it does have a certain amount of pressure drag also, and this can become significant at high angles of attack. The parasite drag of the wing is pretty much dependent on its airfoil, or profile shape, and hence the term profile drag has been attached to it. In some definitions, profile drag is taken to mean the parasite drag of only the two-dimensional airfoil section.

Profile drag does vary with angle of attack or with lift coefficient, as shown in Fig. 3-18 for a typical airfoil. Because this variation is pretty much proportional to the square of lift coefficient, the effect on profile drag by increased angle of attack is usually included in the induced drag term. To account for the variation in all drag with $C_L$, a term known as *Oswald's efficiency factor*, designated as e, is included in the denominator of the induced drag term. The induced drag coefficient is then written:

$$C_{D_i} = \frac{C_L^2}{\pi \times e \times AR}$$

The e accounts for profile drag variation and nonelliptical wing shapes. The drag coefficient included in parasite drag calculation is then the minimum one from the curve, as shown. This same approach can be applied to the drag of the entire airplane and the $C_{D_p}$ value is then referred to as the minimum $C_{D_p}$. This is sometimes written as $C_{D_o}$, which, strictly speaking, means the $C_D$ at $C_L = 0$, but is very close to the minimum $C_{D_p}$, as also shown in Fig. 3- 18.

# SUMMARY OF DRAG DEFINITIONS

It seems that many different labels have been given to identify different types of drag. Hopefully a summary of drag definitions shall avoid too much confusion in this area:

**Pressure drag (form drag).** The drag resulting from an imbalance of pressure forces acting on the surface of a body in the fore and aft directions.

**Skin friction drag (viscous drag).** The drag resulting from the frictional effect of viscous fluid flowing over a surface.

**Parasite drag.** The combined term for all drag except induced drag. Basically parasite drag is made up of pressure and skin friction drag. For a complete airplane, it also includes other drag to which special names have been given.

**Profile drag.** The parasite drag of the wing. Also, sometimes used as the parasite drag of the two-dimensional airfoil section.

**Interference drag.** The additional drag resulting from bringing two bodies together, or in close proximity.

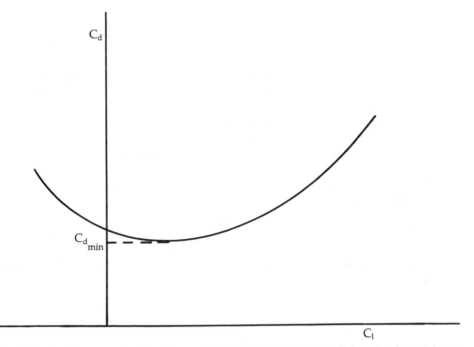

Fig. 3-18. Typical curve of airfoil drag coefficient plotted versus lift coefficient showing minimum drag coefficient.

**Cooling drag.** The drag created by the loss of momentum in the air flowing through the engine compartment for cooling purposes.

**Induced drag.** The drag that results from downwash induced by trailing vortices forming on a lifting surface of finite span.

## GROUND EFFECT

Before leaving the subject of drag, we might also consider a phenomenon that results primarily from drag alteration when flying near to the ground. This phenomenon is appropriately termed *ground effect*. Much dubious information abounds in hangar lore on the reasons for ground effect. Most pilots describe it as "a tendency to float." Some claim that it results from air forced downward by the wing and rebounding against the wing after the air hits the ground. More recently, the popular idea seems to be that a "cushion" of compressed air forms between the wing and the ground. This notion has been promoted by certain companies attempting to hail the merits of their low-wing aircraft.

What really happens when flying close to the ground is a reduction in the downwash of the wing. Recall our discussion on induced drag, which was shown to result from downwash. Downwash, in turn, is caused by the vortex formation at the wingtips. It was pointed out that at high angles of attack, high pressure differential exists between upper and lower wing surfaces, and stronger vortices are induced. High angle of attack in flight is associated with low forward speed. We land and take off at relatively low flight speeds, so that in these operations, there are strong vortices leading to strong downwash and, consequently, high induced drag. This fact can be seen in Fig. 3-15. At lower speeds the total drag is made up more by induced drag and less by parasite drag.

Because induced drag is caused by the vortex action, if vortices could be eliminated, there would be no induced drag. One theoretical (but totally impossible) way of doing this is to have a wing of infinite span. Another way that lies within the realm of possibility is to have a wing with large *endplates*. An endplate is a surface mounted on the wingtip perpendicular to the wing surface. The endplate serves as a barrier to prevent the formation of the vortex. Such devices have been used, but with limited success. In order to destroy, or effectively reduce, the vortex, the endplate would have to be so large that it would have a prohibitive amount of parasite drag itself; thus, it would add more drag than it would eliminate.

However, if the airplane flies close enough to the ground, the ground also serves as a barrier to destroy the vortex action. Because the wingtip cannot be directly against the ground, not all of the vortex is eliminated. Some of it is, though, as shown in Fig. 3-19, and a proportional amount of downwash and induced drag is also eliminated. The closer the wing is to the ground, more of this effect will be felt.

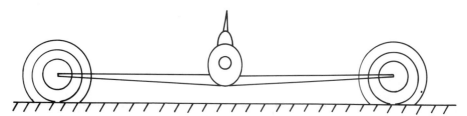

Fig. 3-19. Aircraft in ground effect, showing breakup of tip vortices by ground plane.

Figure 3-20 shows the percentage of normal induced drag (that is, the induced drag far away from the ground) within a spanlength or less above the ground. Above one spanlength, there is essentially no ground effect felt. At approximately 3/10 of a spanlength, only 80 percent of normal induced drag exists, or a reduction of 20 percent. This would represent 9 feet with a wing of 30 feet of span. At 1/10 of a spanlength, the induced drag is reduced approximately 50 percent, but this means only 3 feet of altitude for the same wing.

Obviously, one must be very close to the ground before this effect is significant; therefore, a lot of difference in ground effect will be experienced between a high-wing and a low-wing configuration. Induced drag is a large part of the drag at approach speed, though, and thus even a 10- or 20-percent reduction will have a significant effect on the total drag. With a rather sudden decrease in drag as the ground is approached, the airplane will pick up speed and tend to keep flying much farther down the runway. Lower aspect-ratio-wings also have higher induced drag to begin with, so the effect will be greater with such configurations.

Not all of the floating tendency can be attributed to drag. A reduction in

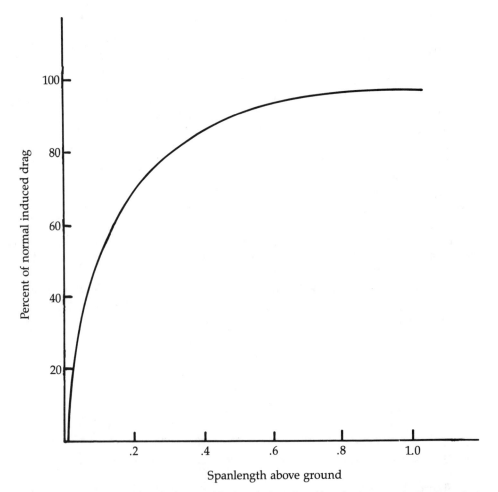

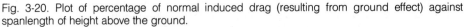

Fig. 3-20. Plot of percentage of normal induced drag (resulting from ground effect) against spanlength of height above the ground.

downwash also changes the orientation of the relative airstream so that the effective angle of attack increases. This is an increase in angle of attack resulting from changing the direction of the airstream rather than the direction of the chordline, as we normally think of it. The effect is the same because angle of attack is the angle between the airstream and the chordline. Increased angle of attack leads to increased lift; therefore, even though it is slight, an additional amount of lift is experienced in ground effect as well as decreased drag. Both effects, of course, act to keep the airplane flying.

A third effect of reduced downwash that works against the pilot's effort to put the airplane on the ground has to do with tail angle. Downwash normally reduces the angle at which the airstream hits the horizontal tail. Decreased downwash thus increases this angle, and gives the tail a higher (or less negative) angle of attack. Such action results in a nose-down pitching motion, also tending to increase speed and reduce drag.

Landing is not the only maneuver in which ground effect plays a part. It is possible to lift off prematurely in ground effect on takeoff. If the aircraft is overloaded and relatively low-powered, it might be possible to break ground, but take an awful lot of additional runway to gain any appreciable altitude. And there is not much ground effect at the standard obstacle height of 50 feet. Airplanes have been known to struggle along for several miles at a few feet of altitude, trying to climb out of ground effect.

Ground effect can be used to advantage, also. It is a way of stretching a glide in the event of a forced landing. Flying close to the ground will also increase range to some extent. The ground must be very flat and open water is probably a more favorable surface over which to practice this trick. Many World War II bombers returned safely to remote Pacific bases by flying at an altitude barely perceptible on the altimeter.

The main reason for this additional flying ability in ground effect is the reduction in induced drag, with a little help from increased lift. Ground effect will have its greatest influence on a low-wing airplane of low aspect ratio at a relatively high gross weight.

# 4

# Thrust

THRUST IS THE FORCE THAT MUST BE GENERATED in order to overcome the natural resistance of drag, just as lift is required to counteract weight. Because drag is the force opposite to the flight path, thrust must be in the direction of flight. With thrust and drag added to it, the force diagram for an airplane appears as shown in Fig. 4-1. For steady, level flight it was shown that lift must just equal weight. In this same case of level, unaccelerated flight, the thrust must also equal the drag. *Unaccelerated* means that the airplane neither speeds up nor slows down, but remains at a constant velocity.

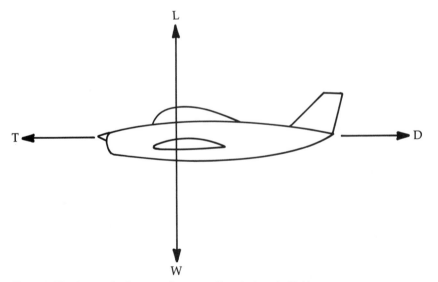

Fig. 4-1. The four major forces acting upon the airplane in flight.

If thrust were increased to greater than the drag, the speed would increase. As the speed increases, so does the drag; it will build up until it equals the new thrust and again the speed will settle out, but at a higher value. The forces of thrust and drag will again be equal and a steady, or *equilibrium*, flight condition will result.

The reverse is true if the thrust were to be reduced. The initially greater drag would decelerate the airplane, but in so doing the drag would decrease. When it reaches a value equal to the thrust, steady flight again results at the lower airspeed. This is a basic principle of physics and applies to any moving vehicle. The thrusting force must equal the retarding forces in order to maintain a steady speed.

Prior to World War II, a discussion of airplane thrust would have been limited to just one power source: the propeller driven by an internal combustion piston engine. Since that great conflict we see many different types of powerplants being used to propel us around our universe. Jets and rockets, which once were identified only with fictional characters like Buck Rogers and Flash Gordon, have now very much become reality. The majority of airplanes are still powered by reciprocating engines driving propellers. The efficiency and low cost of such power has never been surpassed. It will continue for many years to be the major source of light airplane thrust. The reciprocating engine might well be the only type of powerplant that many of us will ever operate. We should, however, have an appreciation for what other types of powerplants can do and why they have come into being.

# PRINCIPLES OF PROPULSION

Before discussing the various ways of creating thrust, we should look at the scientific principles involved. Thrust is a force and force was first examined and explained by Isaac Newton, most renowned for recognizing many principles of science on which much modern technology is based. He determined that a force is equal to mass times acceleration, usually written:

$$F = m \times a$$

In other words, a force exerted on an object of so much mass would accelerate it proportional to the amount of the force. *Acceleration* is a rate of change of velocity or change of velocity over a certain period of time. If a body were initially at rest (have zero velocity) a force would give it some velocity over a certain period of time, and hence give it acceleration. If it were already moving, the velocity would simply be changed from $V_1$ to $V_2$ (Fig. 4-2).

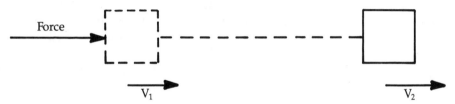

Fig. 4-2. A force acting on a body in motion tends to change it from one velocity to another.

Air has mass, and therefore a force applied to it would tend to accelerate it. A stream of air (or any other fluid) is usually thought of as having a *mass flow*—that is, so much mass flowing per second. A force exerted on a flow of air would increase its velocity (or decrease it, if the force were negative). This action is usually described mathematically as:

$$F = Q \times (V_2 - V_1) = Q \times \Delta V$$

where    Q = mass flow in mass units per second
         $V_1$ = initial velocity in feet per second
         $V_2$ = final velocity in feet per second

$\Delta V$ is just another way of describing a velocity difference or change from $V_1$ to $V_2$.

The above principle is known as Newton's second law. His third law simply states that for every action there is an equal and opposite reaction; therefore, when a mass of air is accelerated, there is an equal and opposite force on the accelerating mechanism. These principles are involved in every type of aircraft powerplant. Air is accelerated by the powerplant in a direction opposite to the path of flight. The equal and opposite reaction is a force on the engine in the direction of flight, and this is the thrust. This reaction principle is easily demonstrated by the movement of a toy balloon as the air escapes during deflation, or in the recoil of a rifle.

Another important principle is also involved in thrust creation. Notice from our equation for the force, which we will now term as thrust, T, that it is proportional to mass flow and velocity change.

$$T = Q \times (V_2 - V_1) = Q \times \Delta V$$

The same amount of thrust could be created with a relatively small mass of air and a large $\Delta V$ or with a relatively large mass and proportionately smaller velocity change; however, it is a fact of life that increasing air velocity imparts higher kinetic energy to the air and some of this kinetic energy is always wasted. It takes engine power to produce this energy, so that the more energy that is wasted, the less efficient the engine is at producing thrust. This fact would seem to indicate that the smaller the change in velocity of the airstream, the more efficient the process would be; however, in order to get a sufficient amount of thrust, a large mass would then be required.

This principle extols the propeller as an efficient thrust producer. A propeller is fairly wide in diameter and takes in a relatively large mass of air per second, as shown in Fig. 4-3. A reasonable amount of thrust can be created with a small change in velocity across the propeller plane and a high degree of efficiency is maintained.

Jet engines, on the other hand, have a relatively small diameter and thus can accommodate a smaller amount of air. In order to produce significant thrust, they must accelerate this small mass to a much larger velocity as shown in Fig. 4-4. In so doing, they waste much kinetic energy, and operate relatively inefficiently.

You might then wonder why we have jets at all. The reason is that they are capable of generating such large velocity changes that they can produce very large amounts of thrust, even though they do so rather inefficiently. Small to moderate amounts of thrust are thus best achieved with propellers that operate at high efficiency. When large amounts of thrust are necessary, efficiency has to be sacrificed for the high thrust capability of the jet engine.

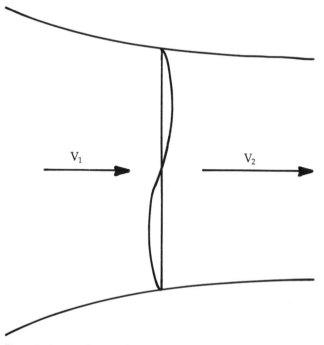

Fig. 4-3. A propeller accelerates a rather large mass of air with a small increase in velocity.

## JET ENGINES

Jet engines are really very simple, in concept, when compared to reciprocating engines. Only the special alloys and critical manufacturing processes required to withstand extreme temperatures make most jet engines very expensive.

The simplest form of jet engine is the *ramjet*. These jets were used on some early missiles and on helicopter rotor blade tips. A ramjet is merely a nozzle-shaped device in which the air is compressed by the ram effect of moving through the air. Fuel is injected and ignited and the expanded gas exhausts at high velocity. They are quite simple and can sustain very high temperatures, thus having the capability for high thrust at very high speeds. Their disadvantage is that they must be in motion at a fairly high speed in order to be started. Normally, this speed is on the order of 300 mph at sea level.

A variation of the ramjet is the *pulsejet*, shown in Fig. 4-5. In this engine, a shutter-like check valve is installed in the air inlet that works in synchronization with the pulsed injection of fuel. The fuel is ignited and expansion forces the

Fig. 4-4. A jet engine accelerates a small mass of air through a large change in velocity.

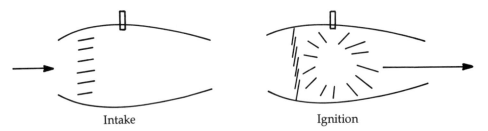

Intake                        Ignition

Fig. 4-5. A pulse jet takes in a certain quantity of air through open shutter valves (A). The valves then close, fuel is injected and ignited (B), causing expanded gases to be ejected at the rear.

check valve shut and a burst of gas out the exhaust. Ram air then forces the check valve open as the internal pressure drops and the cycle is repeated. These engines were made famous (or infamous) when used in the German V-1 "buzz-bombs" of World War II (Fig. 4-6). They too require a forward speed in order to start operation, and were usually launched by means of booster rockets. Neither the ramjet nor its pulsejet cousin have found much application to conventional aircraft propulsion.

Normally, when we refer to the jet engine today, we are speaking of the *turbo-jet*. These engines (and their various modified forms) are used on all contemporary jet aircraft. Turbojets were developed during the 1930s by Sir Frank Whittle of England and Hans von Ohain of Germany, each acting independently and uncognizant of the other. Whittle actually completed the first turbojet and ran it on a test stand in 1937; however, Ohain's engine was the first to fly in the Heinkel He 178 in 1939. Both events aroused little interest from their respective governments. It took the pressures of war to force Germany to try jet propulsion as one of its last desperate attempts to turn the tide of World War II. The effort was too late; jet fighters caused more alarm to enemy aircraft than actual damage. The Messerschmitt Me-262 (Fig. 4-7) was the first jet in battle.

The British eventually flew the Whittle engine in the Gloster E28 and its technology was adapted in the United States first in the Bell P-59 (Fig. 4-8); however, the first production American combat jet was the Lockheed P-80, which appeared in 1944 and never saw action in World War II.

After the war, jet engines were cautiously adapted to commercial transport aircraft. The first of these was the deHavilland Comet in 1949 (Fig. 4-9), which unfortunately suffered from fatigue problems in the fuselage structure. Several disasters with this airplane set back commercial jet development for several years. In the mid '50s, Boeing had more success with the 707 and it was to become the "DC-3 of the jet age."

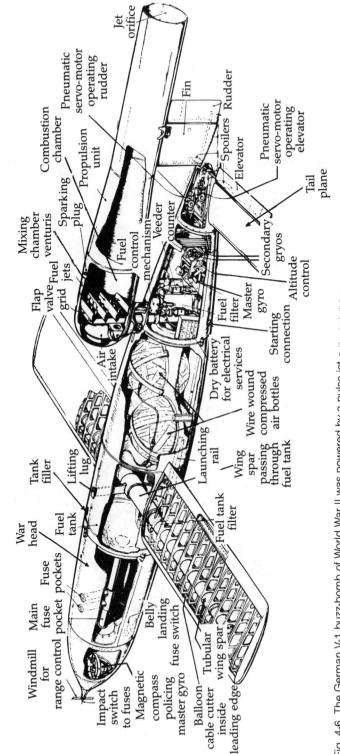

Windmill
for
range control
to fuses

Impact
switch

Main
fuse

Fuse
pocket pockets

Magnetic
compass
policing
master gyro

War
head

Belly
landing
fuse switch

Fuel
tank

Tank
filler

Lifting
lug

Air
intake

Flap
valve

Mixing
chamber
venturis

Fuel
grid jets

Combustion
chamber

Sparking
plug

Propulsion
unit

Pneumatic
servo-motor
operating
rudder

Jet
orifice

Fuel
control
mechanism

Veeder
counter

Fin

Spoilers Rudder

Elevator

Pneumatic
servo-motor
operating
elevator

Secondary
gryos

Altitude
control

Tail
plane

Master
gyro

Fuel
filter

Starting
connection

Dry battery
for electrical
services

Wire wound
compressed
air bottles

Launching
rail

Wing
spar
passing
through
fuel tank

Fuel tank
filter

Balloon
cable cutter
inside
leading edge

Tubular
wing spar

Fig. 4-6. The German V-1 buzz-bomb of World War II was powered by a pulse jet. Smithsonian Institution

Fig. 4-7. The Messerschmitt Me-262, the first operational jet-powered aircraft.

Fig. 4-8. The Bell P-59, the first American jet aircraft.

The engines were developed for military use. As such, they had small intake areas for low drag, which necessitated high exhaust velocities. This design was inefficient and created high noise levels. Skepticism about the feasibility of jet propulsion abounded, even among the experts. Theodore von Karman, one of

Fig. 4-9. The de Havilland Comet was the first commercial jet aircraft, introduced in 1949.

history's greatest aeronautical scientists, wrote in a report in 1940 that he doubted if jet engines could ever be feasibly applied to aircraft. Only a few years after he was proven wrong, K.D. Wood, a well-known aircraft design professor, stated in one of his books that jet propulsion could never be made cheap enough to be used on aircraft with paying passengers.

Nevertheless, the turbojet prevailed. It differs from the simpler types of jet engines by the fact that it has a compressor to compress the incoming air. This enables the turbojet to start from zero velocity without auxiliary launching devices.

Figure 4-10 shows a schematic of an *axial-flow* turbojet. Incoming air is compressed by the compressor and forced into the burner section. Fuel is injected here and ignited. The burning gases rapidly expand with this additional energy added and force their way out through the exhaust (incoming compressed air prevents exit in the forward direction). As the exhaust gas exits, it passes through the turbine, giving it rotational motion. The turbine is connected to the compressor through a shaft and drives the compressor to bring in more air. The remainder of the gases are exhausted. As they expand, the pressure drops to atmospheric and is traded off for high velocity. The reaction to the acceleration of this mass of air is a thrusting force against the engine.

Such engines are usually referred to as *pure turbojet* engines. In pure jets, all of the air flows through the combustion section and gets accelerated to high velocity. The inefficiency of this configuration was recognized, but not considered too significant in the early days of jet travel. As their use increased, it became apparent that the old military jet designs were no longer suitable for profit-making aircraft. Designers began to design engines specifically for commercial use. To increase the efficiency but retain most of the high-thrust ability of the turbojet, engineers developed the *turbofan*.

A turbofan is essentially a turbojet with a ducted fan mounted on the shaft,

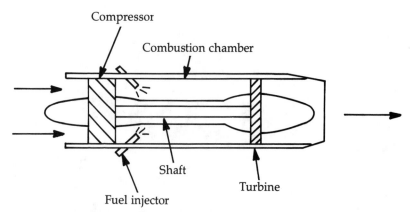

Fig. 4-10. Diagram of an axial-flow turbojet engine.

usually ahead of the compressor as shown in Fig. 4-11. The fan serves as a first stage compressor for air going through the combustion section. It also accelerates a large mass of air that does not pass through the burners, but goes around the engine. This bypass air does not get burned and is thus accelerated to a lesser degree than the gases emitted from the burners. It therefore loses less kinetic energy and is a more efficient thrust producer.

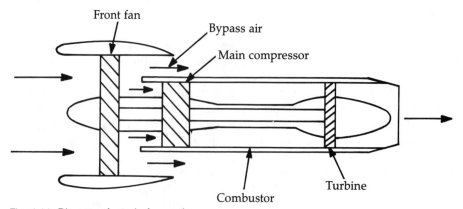

Fig. 4-11. Diagram of a turbofan engine.

The turbofan has the advantage of improved efficiency over a turbojet. The bypass air from the fan provides this increased efficiency. It also derives part of its thrust from the jet engine section, which, of course, accelerates the air to a higher degree, and obtains a significant amount of thrust. The turbofan is a compromise between a pure jet and a propeller propulsion system.

The ratio of the mass of air passing around the engine to that passing through the engine is termed the *bypass ratio*. The larger the bypass ratio, the more efficient this powerplant becomes; however, the maximum thrust capability also decreases with this ratio increase. The turbine must drive the compressor and the fan. For very large fans, much of the exhaust gas energy is absorbed in powering the fan, leaving less for jet thrust.

The ultimate extension of the turbofan is to replace the fan and its duct with a propeller, and to utilize most of the engine power to drive the propeller. Such an arrangement is called a *turboprop* and is shown in Fig. 4-12. The turbine in this engine drives the compressor and the propeller. Because jets (or *turbine engines*, as they are more properly called) run at very high rpm, the shaft rotation must be slowed through reduction gearing before being attached to the propeller. Although most of the turboprop engine's output is through the propeller, a small amount of jet thrust can be obtained from the exhaust. This is usually approximately 10 to 15 percent of the total thrust, with the rest coming from the propeller. Turbine engines are more costly than reciprocating engines and use more fuel; however, they have a high power output for their size and weight. For this reason the turboprop has been enjoying increasing popularity in higher performance light aircraft.

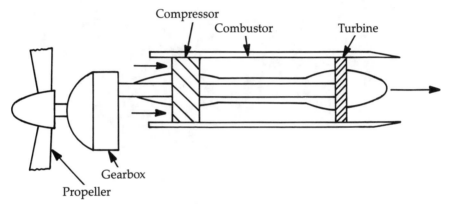

Fig. 4-12. Diagram of a turboprop engine.

# RECIPROCATING ENGINES

The *internal combustion reciprocating* (piston) engine was the first engine to successfully power an airplane. For 40 years it remained the only airplane engine. Even today, it is still the most popular aircraft powerplant, with its relatively low cost and high degree of efficiency. Internal combustion engines were developed in the mid-1800s and the four-stroke cycle engine, the version used in aircraft engines, is credited to a German named Nickolaus Otto in 1876.

At the time the Wrights, Langley, and others were looking to powered flight, this engine appeared to be the only one capable of producing sufficient power with reasonably low weight. Steam engines had been tried, but unsuccessfully, because they were heavy and cumbersome. Power-to-weight ratio has always been the hallmark of a good aircraft engine. By 1900, there had not been much effort made to develop engines with proper power-to-weight ratios.

Samuel Langley, whose flying machine failed to fly just nine days before the

Wrights' first airplane flight, actually possessed a better engine. His engine, built by Charles Manley and finished in 1902, produced 52 horsepower and weighed 200 pounds. The first Wright engine developed 12 horsepower and weighed 100 pounds, about half the power-to-weight ratio of Langley's. Nevertheless, the Wrights' engine was a notable achievement. Coupled with their attention to many other aspects of flight, it provided them the ability to achieve that remarkable feat at Kill Devil Hill that gave us a new dimension to our world (Fig. 4-13).

Fig. 4-13. The Wright brothers' engine that powered the first Wright Flyer.

The reciprocating engine is a terribly complex device when compared with a turbine engine. The first downward stroke of the piston is timed to coincide with an intake valve opening into the cylinder. Fuel-air mixture is drawn in, which has been carefully combined in the proper ratio by a carburetor (a complex device in itself). The piston then makes an upward stroke, compressing the gas, and just at the right time a spark must be discharged by a totally separate ignition system.

The burning then drives the piston down in the only one of the four strokes that really provides any power to the crankshaft. A fourth stroke (upward) is necessary to exhaust the remaining gas, also timed with the opening of the exhaust valve. Power has now been supplied to rotate a shaft running parallel to the aircraft longitudinal axis. The rotating shaft turns a propeller at one end, which provides a thrust force in the forward direction.

Amazingly, it works. Because it was the only powerplant for airplanes for 40 years (and the only one for commercial craft for even longer), it has been continually refined to the point where it works very well.

The first reciprocating engines were liquid-cooled. The cooling system adds weight, but the first airplanes were so slow that air-cooling was not practical.

This problem was solved a few years later with rotary engines (Fig. 4-14). The crankshaft of a rotary is rigidly mounted to the airframe, and the cylinders revolve around it. In this way the cylinders are cooled by air flowing over them, even at low speeds.

Smithsonian Institution

Fig. 4-14. A Gnome Rotary engine that powered many early World War I fighters.

Eventually, stationary-cylinder air-cooled engines proved successful by use of adequate cooling fans and with increased airspeeds of newer designs. The liquid-cooled engine was used for many more years in certain types of aircraft. The low frontal area of in-line engines allowed greater speed to be developed by racing and fighter aircraft in which they were installed. In-line arrangements were difficult to adequately cool with air, so liquid-cooling worked better in this case.

Air-cooled engines first employed cylinders placed radially around the crankshaft. This arrangement allowed for a good number of cylinders, providing much power, but with all of them having ample exposure to cooling air. The disadvantage in this arrangement was the high drag resulting from the large frontal area. Less drag was encountered with in-line arrangements but only a few cylinders could be so arranged and adequately cooled.

A horizontally opposed configuration worked very well for low-power engines, when only two or four cylinders were required. This arrangement lent itself very well to efficient cowling design for light airplanes, and enjoyed rapidly increasing popularity following World War II. As the requirement for large, high-power engines began to be filled by turboprops, radials disappeared. Light airplane engine manufacturers zeroed in on the horizontally opposed type, which has come to be the mainstay of lightplane power. Horizontally opposed reciprocating engines have been developed with as many as eight cylinders, and producing up to 450 horsepower.

# RECIPROCATING ENGINE PERFORMANCE

The power output of reciprocating engines is measured in English units in *horsepower*. One horsepower is equal to 550 foot-pounds per second. In other words, it would take one horsepower to move something requiring a force of 550 pounds over a distance of one foot every second. In metric units, the term for power is *kilowatts*. Although metric conversion will probably come to pass in the United States, currently the light aircraft industry uses English units, and indications are that it will for some time to come. Engine power is thus always found for American engines in terms of horsepower.

The capability of a cylinder of an engine to develop power depends on a number of variables. The force on the piston in the power stroke is equal to the pressure of the burning gas exerted against the piston times the area of the piston. Because this pressure varies throughout the stroke, a *mean effective pressure*, or average pressure, is considered. The distance that this force moves the piston determines the work done on the piston. The distance that it moves is, of course, the length of the stroke.

Power is really the time rate at which work is done. The number of power strokes per minute is $1/2$ times the rpm because there is a power stroke every other revolution; therefore, the power in one cylinder is proportional to average cylinder pressure times length of stroke times piston area times rpm. This is written as:

$$\text{Power} \propto \text{PLAN}$$

where   P = mean pressure.
        L = length of stroke.
        A = cross-sectional piston area.
        N = engine speed or rpm.

Power for the total engine would then be equal to that of one cylinder times the number of cylinders. Sometimes the formula for a complete engine is written as:

$$\text{Power} \propto \text{PLAN}_n$$

where the small n stands for number of cylinders.

The power formula reveals some very basic qualities of the piston engine. Length of stroke times piston area (L × A) is *displacement* of the cylinder (essentially, its volume); therefore, power for a given basic engine design (one with a given displacement and number of cylinders) could only be increased by increasing either the engine rpm (N in the formula) or the pressure in the cylinders (P).

The air in the cylinders is compressed in proportion to the compression ratio, but the final pressure depends on how much pressure the air had at the start of compression. This pressure is the pressure of the air in the intake manifold. Engine power is thus controlled by varying either the intake manifold air pressure (MAP) or the rpm, or both. The power of the engine could be increased if the basic design were modified either to increase the compression ratio (which also increases mean effective pressure, P) or to increase the displacement.

Not all of the internal power in the cylinders is delivered at the output shaft of the engine. Some power is lost in the friction of the moving parts. The horsepower delivered at the shaft is termed *brake horsepower*, or bhp. It is so named because the classical method of measuring power output was by use of a device known as a *Prony brake*.

If the engine is connected to the propeller through reduction gears, additional power is lost through the friction of the gears. Gearing is sometimes utilized to get more power out of an engine design by running at very high rpm; however, the propeller is limited in speed as its blade tips approach the speed of sound. At such high speeds, very high drag can be encountered, so this situation must be avoided. Because the power at the propeller shaft can be different from bhp, the term *shaft horsepower* (shp) represents the power delivered to the propeller. For direct drive engines, bhp and shp are one and the same.

Another term, which we will use in more detail later, is *thrust horsepower* (thp), which is the amount of power that actually gets converted into thrust. Some of the power delivered to the propeller is lost in drag, the swirl of the slipstream, and other energy-robbing devices. The amount of the bhp (for direct drive) that can be converted to thp is a measure of the efficiency of the propeller; thp is thus defined as the propeller efficiency times the brake horsepower:

$$thp = \eta_p \times bhp$$

where $\eta_p$ is the propulsive efficiency.

Because the engine power is proportional to the pressure of the air in the intake manifold, and this air comes from the atmosphere, power is dependent on atmospheric pressure. Pressure or density of the air is highest at sea level and decreases with altitude. The engine therefore develops its greatest power at sea level. At higher altitudes, the power output of the engine is proportionately less. Figure 4-15 shows how the brake horsepower of a typical engine varies with altitude for any given value of rpm.

Engines are designed to run at a certain maximum value of rpm. The power output at this rpm at standard sea level density is called the *rated brake horsepower*. The engine depicted in Fig. 4-15 develops 200 horsepower at sea level at 2,700 rpm. This would be its rated bhp if this were the maximum design rpm. This figure also shows that at 5,000 feet, even at full rpm, only 165 bhp are available. As

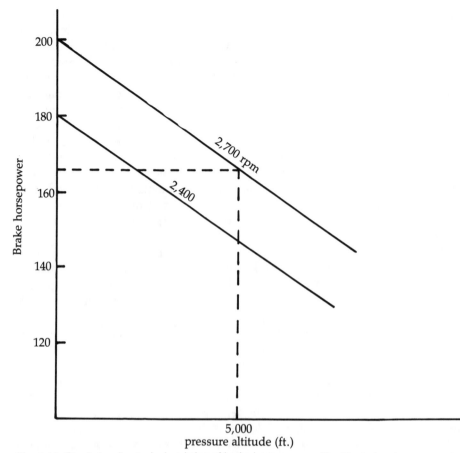

Fig. 4-15. Chart showing typical variation of brake horsepower with altitude for given rpm values.

we go higher, the engine develops even less power. Of course, at lower rpm at any given altitude, it also develops less power.

Most piston engines can be run at full rated power for a considerable amount of time without immediate damage; however, their life would be greatly reduced and fuel efficiency would be extremely poor. Usually, the maximum recommended power for continuous cruise is 75 percent of the rated bhp. This amount of power is a fixed value for each engine; 75 percent of 200 horsepower, for example, is 150 horsepower; therefore, operating at 75 percent power means at 150 horsepower with this engine, not 75 percent of the maximum power that is available at a certain altitude.

Running at too low a power setting can foul spark plugs and valves and is also not recommended. Usually, the normal range of operation is considered to be between 55 percent and 75 percent. Fifty-five percent power is sometimes regarded as *economy cruise* setting because fuel consumption is proportional to power. Seventy-five percent power, on the other hand, consumes more fuel, but gives better performance, and is sometimes referred to as *performance cruise power*. The intermediate position of 65 percent is a good compromise, combining reasonable performance with reasonable economy.

# SUPERCHARGERS

High altitude is a regime where efficient flight is possible because the lower density of the air causes low parasite drag. The lower density, on the other hand, deteriorates engine performance, so that the net result is usually a decrease in performance rather than an increase. This situation could be greatly improved if higher pressure air could be supplied to the intake manifold. One way of doing this is with a *supercharger*.

Superchargers utilize a small compressor in the intake manifold that compresses the air received from the atmosphere to a higher pressure. Originally the compressor was driven by a drive tapped off of the engine crankshaft through a gear train. This arrangement takes some power from the engine to power the supercharger. A more efficient way of powering the supercharger is by a turbine placed in the exhaust flow. This device is known as a *turbosupercharger*, or simply, *turbocharger*.

Turbosuperchargers are more efficient because they use exhaust gas pressure, which is wasted energy, and do not require as much power from the engine. The exhaust drives the turbine, which is connected to the compressor in the intake, much like the compressor and turbine function in a turbojet. Of course some back pressure is created by the turbine, which decreases engine power; however, the net gain is significant. Turbosuperchargers are also capable of a greater degree of supercharging than the gear-driven type. Figure 4-16 shows a typical turbosupercharger.

The effect of supercharging (or *turbocharging*, as it is now called) is that sea level, or rated, power can be maintained up to a certain altitude. The altitude above which sea level power cannot be maintained is known as the *critical altitude* and is on the order of 7,000 or 8,000 feet for current designs. Figure 4-17 shows the effect of supercharging on our 200-horsepower engine used in the previous example. The dotted line represents the power available at a constant rpm with turbocharging.

Notice that the 200 horsepower is constant for considerable altitude and then drops off from the critical altitude point. The solid line represents the original unsupercharged engine. It should be obvious that greater power is possible with the turbocharger at altitudes below and above the critical altitude. Turbochargers have become increasingly popular in recent years as fuel costs and shortages have required more efficient methods of flight. Even many single-engine airplanes are available in turbocharged versions.

# PROPELLERS

The reciprocating engine with all of its improvements and refinements over the years cannot produce thrust by itself. It can only drive a device that actually does perform this job. The device used, of course, is the *propeller*. The propeller might

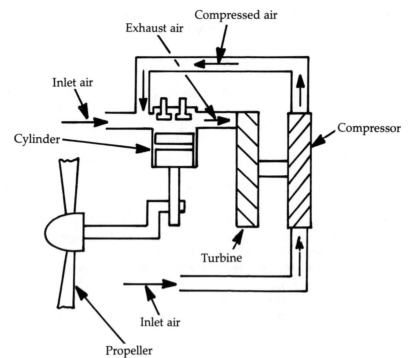

Fig. 4-16. Diagram of a turbosupercharger.

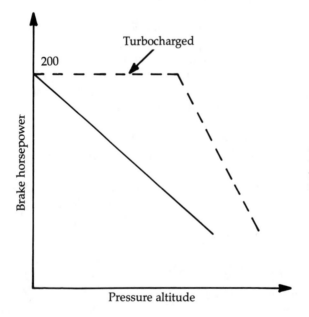

Fig. 4-17. Chart showing the change in brake horsepower variation with altitude when a turbocharger is installed.

seem like a simple piece of hardware, but it is rather complex. When it was first conceived, it was felt that a propeller would screw through the air much as a screw moves through wood. The term *airscrew* is used by the British, and marine propellers are still referred to as screws.

Propellers as a means of providing aircraft thrust were suggested as far back in history as Leonardo da Vinci. They were developed for boats in the early 1800s and applied to airships in the mid-1800s; however, early propellers were grossly misunderstood and consequently very inefficient.

Again, it took the Wright Brothers to realize that most existing propeller theory was all wrong and that research and experimentation had to be done in order to develop a propeller that could successfully propel a powered airplane. They performed this task and it is just one more of their notable achievements that made powered flight possible. They were really the first to recognize that a propeller is essentially a small wing rotated in a plane perpendicular to the path of flight and developing thrust in the same way that a wing develops lift.

Figure 4-18 illustrates the manner in which the propeller functions like a wing. A slightly exaggerated propeller blade is shown, looking down at its tip from above the nose of an airplane. The propeller is being rotated with the airplane standing still. The relative wind strikes the blade in the plane of rotation just as it does a wing moving through it. The blade is set at an angle of attack, and because it has a cambered airfoil cross-section, it develops lift in the forward direction. A force in this direction is normally referred to as *thrust*.

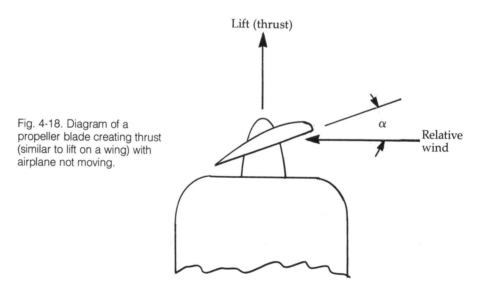

Fig. 4-18. Diagram of a propeller blade creating thrust (similar to lift on a wing) with airplane not moving.

Now suppose that the airplane starts moving forward. The blade now has a forward motion as well as rotational motion. Figure 4-19 shows a cross section of the blade in this situation. The velocities of the blade itself are shown for clarity rather than the relative wind vectors, which would be the same, except in opposite directions. The blade is seen to have a component of speed tangential to the rotating path (at any given point along this path) called, simply, the *rotational velocity*, $V_{rot}$, and a component of *forward speed*, $V_{fwd}$. The resultant of these velocity vectors is then labeled $V_{res}$. This means that the blade is actually moving, at that point, along the path of the resultant velocity and the relative wind is meeting it along that path in the opposite direction.

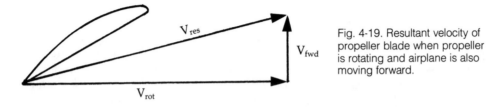

Fig. 4-19. Resultant velocity of propeller blade when propeller is rotating and airplane is also moving forward.

The angle of attack for an airfoil is the angle between the relative wind (or resultant velocity) and the chordline. The angle of attack for our blade section is this same angle and is shown in Fig. 4-20 as $\alpha$. Notice that with forward motion, the angle of attack is not the same as the angle between the chordline and the plane of rotation. This angle is really the *pitch* angle of the blade. Pitch angle and angle of attack are only the same if no forward motion is involved.

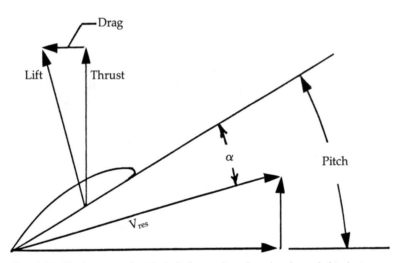

Fig. 4-20. Moving propeller blade in forward motion showing relation between pitch angle and angle of attack.

Also recall that lift was the force created on an airfoil perpendicular to the relative wind. The lift vector of the propeller, then, with forward motion, does not point exactly forward; therefore; not all of it contributes to thrust. Thrust is only the forward component of the lift. Some of the lift is now working against the motion of the blade. This component adds to the drag of the blade and takes power from the engine to overcome. This force along with the other drag of the blade (parasite and induced) accounts for the loss in power as the engine power is converted to thrust power.

So far we have been discussing the blade at one location out along its radius. Actually, each point along the radius has a different tangential velocity for the same rotational speed (same rpm). Figure 4-21 shows the variation in velocity for different radius points, or distance out from the center of the hub. This is the same as the velocity variation along the radius of a wheel.

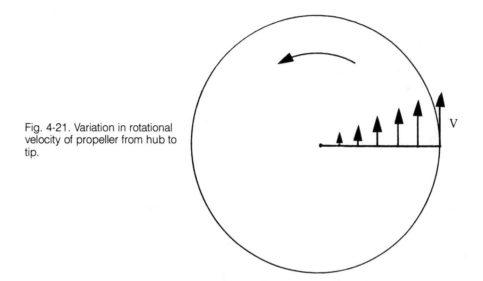

Fig. 4-21. Variation in rotational velocity of propeller from hub to tip.

The vector diagram of Fig. 4-19 is different, then, for each station, and with the same forward velocity, the resultant velocities end up at different angles to the perpendicular. This situation is portrayed in Fig. 4-22. To keep the angle of attack the same along the radius, the blade has to be twisted. It has a high pitch angle near the hub and a relatively low angle toward the tip. All propellers are twisted like this, as can be seen in the photo in Fig. 4-23. It is interesting to note that the Wrights were the first to recognize this requirement in an efficient propeller blade. This can be seen in their propeller, shown in Fig. 4-24.

Because the propeller blade section acts just like a section of a wing, let us review some basic airfoil behavior. Recall from chapter 2 that lift increases in direct proportion to angle of attack; however, there is also some angle of attack (or lift coefficient) where the drag is a minimum, as seen in Fig. 3-15. Above and below this angle of attack, the drag increases; therefore, the ratio of lift to drag has a maximum point corresponding to the optimum angle of attack, or $\alpha_{opt}$, and becomes less above and below this angle, as shown in Fig. 4-25. For a propeller, the most lift for the least amount of drag means the most thrust for the least amount of power lost; therefore, operating the propeller blade at its optimum angle of attack gives yields greatest efficiency.

Suppose that the angle of attack shown in Fig. 4-20 was the optimum for this

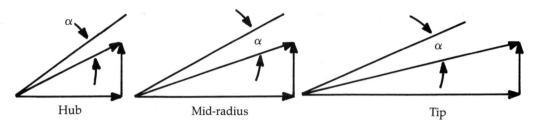

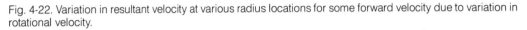

Fig. 4-22. Variation in resultant velocity at various radius locations for some forward velocity due to variation in rotational velocity.

Fig. 4-23. Twist built into propeller blade to maintain the same angle of attack at each radius location.

airfoil. (Remember, the angle of attack depends on the orientation of the resultant velocity, which also depends on forward and rotational velocity.) Next, maintain rpm so the rotational velocity remains the same, but increase the forward speed. This situation is depicted in Fig. 4-26. The original resultant velocity (indicated by the solid line) has moved forward by the amount indicated by the dotted line, resulting in a decreased angle of attack. Drag is thus decreased, but so is lift, actually more so, and the resulting L/D ratio (the measure of efficiency) is less.

Conversely, a decrease in forward velocity will increase angle of attack, as shown in Fig. 4-27, by moving the resultant velocity vector farther aft. In this case, drag greatly increases, and although lift increases somewhat, the net L/D ratio goes down. Of course, the same results would occur if forward speed were kept constant but the rotational velocity vector changed by increasing or decreasing rpm.

Fig. 4-24. The Wright brothers' propeller, the first to have twist to maintain constant angle of attack throughout the radius.

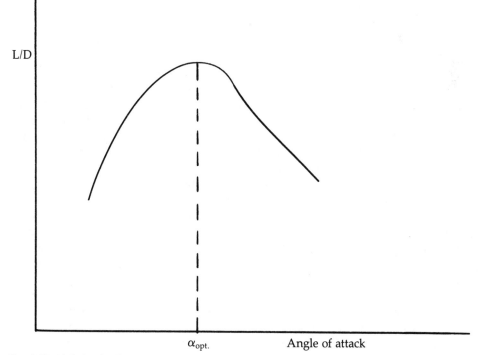

Fig 4-25. Variation in lift-to-drag ratio with angle of attack of an airfoil showing optimum angle for maximum L/D.

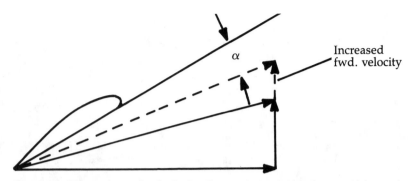

Fig. 4-26. Change in angle of attack with constant rpm but increased forward velocity.

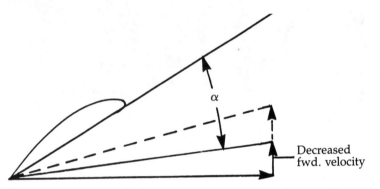

Fig. 4-27. Change in angle of attack with constant rpm but decreased forward velocity.

The efficiency of the propeller, which is really how much brake horsepower it converts into thrust horsepower, is obviously dependent on the ratio of forward speed to rotational speed. Figure 4-28 shows the relationship of efficiency to this ratio for a typical propeller. The maximum that can usually be obtained is approximately 80 to 85 percent. For an installed propeller, 80 percent is usually a more realistic figure to actually count on.

The efficiency curve in Fig. 4-28 represents the efficiency at one pitch angle—or, in other words, depicts the situation for a fixed pitch propeller. Depending on the operating velocity of the airplane and the corresponding rpm, a propeller of appropriate pitch can be selected so as to place the peak efficiency at the ratio of these values. Usually the condition selected is cruise, which involves a relatively high pitch. Sometimes, however, the primary mission of an airplane is to climb, such as for a glider-tow craft.

In this case, a low speed and high rpm moves the velocity ratio to the low side of the chart. A low-pitch propeller will move the peak of the efficiency curve in this direction and give greatest performance in this area. Metal propellers can actually be twisted to different pitch settings by the factory or appropriate facilities. Propellers optimized for cruise condition are sometimes called *cruise props,* and those pitched for climb, *climb props.*

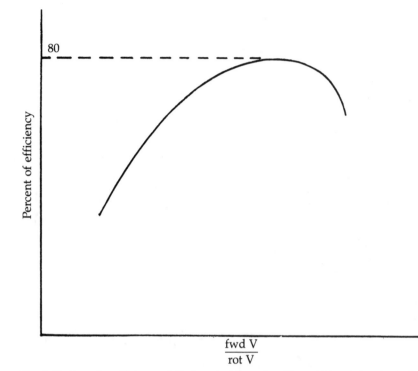

80

Percent of efficiency

$$\frac{\text{fwd V}}{\text{rot V}}$$

Fig. 4-28. Propeller efficiency plotted against the ratio of forward to rotational velocity.

With a propeller of fixed pitch, you only have one condition where the propeller works best. In most flight operations, the airplane must perform at a variety of speeds and power settings. If these operations encompass a wide range of speeds, as they would with an airplane of high cruise and low stall speed, the fixed-pitch propeller is not very practical. The solution, then, is to incorporate a propeller of variable pitch.

Figure 4-29 shows what happens in the case of a *variable pitch propeller*. The dotted lines indicate the position and resultant velocity of our original example propeller. Suppose again that forward velocity is reduced (keeping rotational velocity the same), but that pitch control is capable of reducing our pitch angle. The pitch is reduced so that the original angle of attack—and corresponding efficiency—is maintained. We could now climb with the propeller at peak efficiency and subsequently increase pitch to match the peak efficiency to cruise condition when we level off.

The advantage of the variable pitch is obvious. What results is an efficiency curve as shown in Fig. 4-30. An infinite number of curves of various individual pitch angles is available. It is just like having a lot of propellers of different pitch that can be instantly changed in flight. The peak efficiency can be shifted to any desired ratio of forward to rotational speed.

In actual practice, the pilot has no knowledge of the best pitch angle to select, nor the pitch at which the propeller is set. When pitch is changed, the increase or decrease in drag reduces or increases the rpm accordingly; therefore, rpm is a

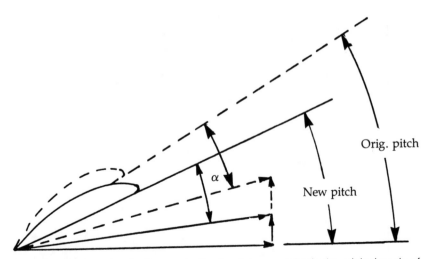

Fig. 4-29. Diagram showing how reduction in pitch can maintain the original angle of attack if forward speed is reduced.

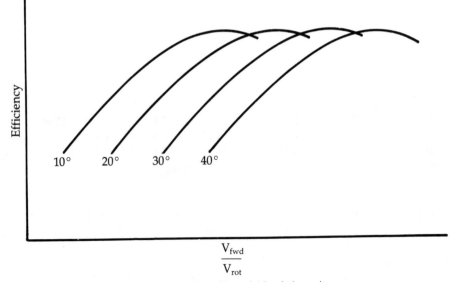

Efficiency

10°    20°    30°    40°

$$\frac{V_{fwd}}{V_{rot}}$$

Fig. 4-30. Efficiency curves for a propeller with variable pitch angles.

measurement of the propeller setting with a variable pitch propeller.

In this case, the throttle adjusts manifold pressure. Because power settings (certain percentages of rated horsepower) are obtained by proper combinations of MAP and rpm, it is desirable to adjust these values and have them stay as set (sort of a set-and-forget operation). To provide this feature, modern variable pitch propellers are *constant-speed*.

Constant-speed propellers incorporate a governor that is adjusted to a certain rpm by the propeller control. The governor then maintains that rpm by adjusting the blade angles accordingly. If the airplane's forward speed increases, the propeller would tend to speed up, but the governor increases the pitch angle to prevent this. If the plane slows down, the propeller would see an increased drag and tend to slow down. Again, the governor responds and reduces blade angle. In this way, the desired rpm is maintained, even though forward speed fluctuates. Operating in turbulent air with a variable pitch propeller without the constant-speed feature, a pilot could be a very busy person.

# TURBOPROP AND TURBOJET PERFORMANCE

We usually think of propellers as being associated with reciprocating engines; however, they can also be powered by turbine engines, as previously mentioned. While the turboprop (propeller-turbine engine combination) is not quite as efficient as the reciprocating engine, it is a great improvement over the turbojet. Considerable power is available in turbines, even very small ones, and this fact makes them desirable in larger aircraft. For aircraft that must operate from small airports, the propeller also gives them relatively high thrust at low speeds, and the turboprop thus becomes the ideal powerplant.

Recall from the discussion regarding reciprocating engine performance that the brake horsepower output at a given rpm is essentially constant for any particular altitude. This situation is different with turbine engines. At higher airspeeds, the ram effect of the air increases compression of the intake air and increases the power output. As in the case of reciprocating engines, power for turboprops also decreases with altitude because the power depends on air density. Figure 4-31 shows the variation of power available for a typical turboprop engine with airspeed and altitude. Because turboprop engines are always geared down to propeller shaft speed, the output is measured in *shaft horsepower* (shp) rather than bhp, as with reciprocating engines.

This chart also shows a constant shp at sea level. This indicates that although the engine might be capable of developing higher power, it is limited to no more than the rated power. This is because the engine is only designed to operate at this value of power; operating at higher output could overstress or overheat certain engine parts. Such limitation is often referred to as flat rating, meaning a constant power (appearing flat on the chart) up to a certain speed or altitude. Many turbine engines have such restrictions; however, this practice is not limited to these engines and might apply to any type of engine, particularly those in the higher performance categories.

Most turboprop engines also develop a small amount of jet thrust. This thrust is independent of thrust obtained from the propeller. Because most of the thrust comes from the propeller, turboprops are normally rated in horsepower. To account for the additional jet thrust, it is converted into a power term and added to the shaft horsepower to yield a term known as *equivalent shaft horsepower* (eshp).

$$eshp = shp + \text{power from jet thrust}$$

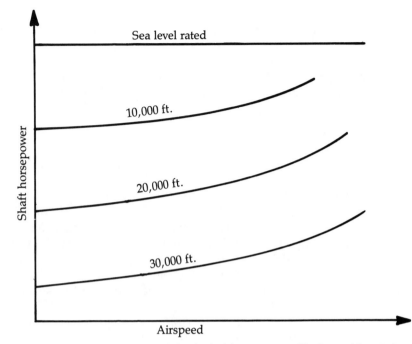

Fig. 4-31. Chart showing the variation in shaft horsepower with airspeed for a turbo-prop at various altitudes.

Pure turbojet engines are normally rated not in terms of power, but in terms of their thrust output. Like all air-breathing engines, the thrust of a jet decreases with altitude. The variation in thrust with airspeed is not easily defined. The performance of a turbine engine depends upon a large number of variables such as air temperature and pressure, compressor ratio and temperature, turbine temperature and pressure, exhaust velocity and temperature, and the like.

Most jet engines display thrust variation somewhat like that shown in Fig. 4-32. The thrust drops off slightly as speed increases from zero and then increases again at higher speeds. For the purpose of estimating jet thrust, a constant value for a given altitude is often assumed. This method works in many applications where a first estimate must be made to determine the approximate performance of an aircraft.

## COMPARISON OF POWERPLANT PERFORMANCE

This chapter has continuously emphasized that different types of propulsion are best for different flight conditions. One major reason for this fact is the variation in propulsive efficiency with airspeed; greater efficiency results when a larger mass of air is accelerated over a lower increase in velocity; therefore, the propeller turns out to be the winner in the efficiency race and really stands out at low airspeeds. This fact is dramatically illustrated in Fig. 4-33, which illustrates the efficiency of various propulsion systems at various airspeeds.

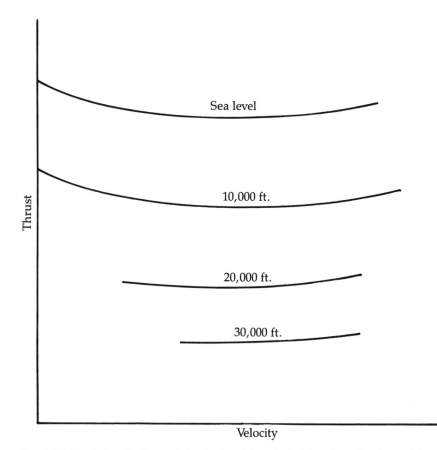

Fig. 4-32. Chart showing the variation in thrust for a turbojet engine with airspeed for various altitudes.

The propeller exceeds the turbofan and turbojet at speeds lower than approximately 350 mph; however, once that speed is exceeded, propeller efficiency drops off rapidly because at these speeds, propeller blades have to travel so fast that they approach the speed of sound and shock waves can form. When this happens, drag increases rapidly, and much engine power is lost in overcoming this drag.

Beyond 400 mph, the turbofan moves ahead of the pack in efficiency. This curve is actually for a relatively high-bypass ratio turbofan; therefore, an upper speed limit also exists due to limitations on blade speed. The area of the turbofan, which is smaller than the propeller but much larger than the turbojet, gives the turbofan an edge up to approximately 900 mph, with peak efficiency at 700 mph. This range includes the cruise area of subsonic airliners and business jets, and consequently appears to be the best engine for such aircraft.

Only at supersonic speeds does the pure turbojet take over the lead in efficiency, which continues to increase well off of this chart. Efficiency does peak at approximately 2,500 mph, but this is due mostly to a temperature barrier that

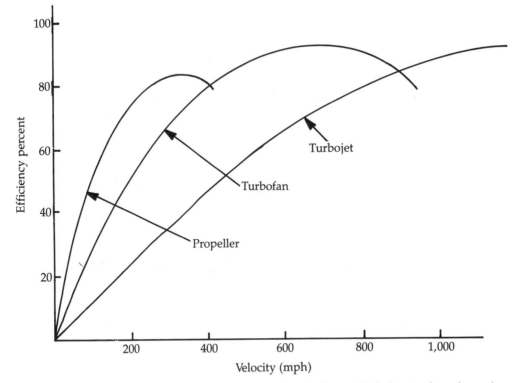

Fig. 4-33. Chart showing the relative efficiency of propeller, turbofan, and turbojet at various airspeeds.

exists in that regime. Turbines are not capable of withstanding the extreme gas temperature required above such speed.

The actual amount of thrust that can be developed by a turbine engine in various configurations is shown in Fig. 4-34. Here the same basic engine is compared when it functions as a pure turbojet, when it is converted to a turbofan, and when it is used to drive a propeller. Clearly, the greatest thrust is available at low airspeed with the turboprop. The top speed, though, is limited to about 400 mph. The turbojet has lowest thrust at low speed. The turbofan appears to be a good compromise, giving relatively high thrust at low and higher speeds.

Another way of considering the relative efficiency of an engine, particularly nowadays, is the amount of fuel it consumes in producing thrust. The measure of fuel efficiency is the term *specific fuel consumption* (SFC). When considering jets or turbine engines, the term *thrust specific fuel consumption* (TSFC) is normally used. Thrust specific fuel consumption refers to the amount of fuel that is required for an engine to develop an equal amount of thrust in a given period of time. In English units this is taken as pounds of fuel per pound of thrust per hour.

$$TSFC = lbs. \ fuel/lbs. \ thrust/hour$$

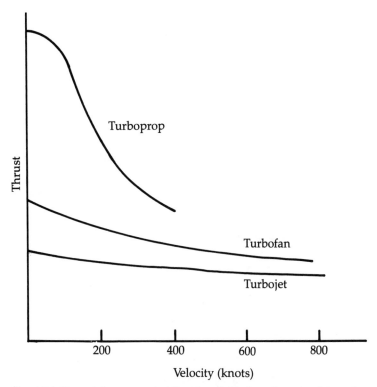

Fig. 4-34. The relative amount of thrust available from the same jet engine when used as a pure jet, converted to a turbofan, or converted to a turbo-prop.

Figure 4-35 shows the relative fuel consumption of the three engines compared in Fig. 4-34. Again, note the greater efficiency of the turboprop in this area by its lowest TSFC, particularly at very low speed. As might be expected, the turbojet loses this race with the highest overall fuel consumption, and the turbofan comes out in the middle.

Reciprocating engines are normally rated for fuel consumption by how much fuel is required per hour to develop a unit of power (rather than thrust); therefore, American engines are rated in pounds of fuel per brake horsepower developed per hour. This value is called simply *specific fuel consumption* (SFC). Unless the T is included, the fuel consumption is implied to be in terms of *power* produced:

$$SFC = \text{lbs. fuel/bhp/hour}$$

Fuel consumption is normally specified in pounds of fuel rather than gallons. This is because the energy content of fuel is proportional to its weight. Actual volume of a given weight of fuel can vary depending on the temperature, and the temperature variation over the flight regime of many aircraft can be quite large.

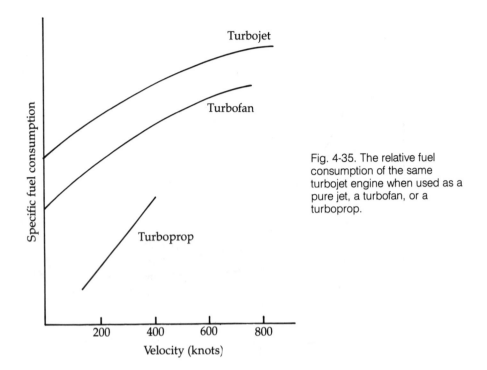

Fig. 4-35. The relative fuel consumption of the same turbojet engine when used as a pure jet, a turbofan, or a turboprop.

Modern reciprocating engines have specific fuel consumption values on the order of 0.4 to 0.5 lbs/bhp/hr., with 0.45 usually considered the average.

Turboprop engines, when rated in SFC, usually consume approximately 0.6 lbs/eshp/hr., although recent developments in turboprops are improving this figure; however, at present, reciprocating engines driving propellers represent the best overall combination of an efficient propulsion system. If we can tolerate speeds of a few hundred knots, this seems to be the way to go.

# 5

# Performance

COMBINED EFFECTS OF DRAG AND THRUST, which are examined in chapters 3 and 4, pretty much determine the airplane's *performance*. Performance means how fast it will go, how fast it will climb, how quickly it will take off and land, and how far it will go. There are other desirable virtues of flying machines, but these items are usually of foremost concern. This chapter examines how the forces on the airplane affect its behavior in these areas. In general, almost anyone could reason that lower drag and higher thrust would improve performance. But there are limits on what is possible, and the degree to which performance improves with changes in airplane characteristics is not quite so obvious.

## LEVEL FLIGHT PERFORMANCE

Probably the foremost advantage of air travel is its speed. Most people fly (either as pilots or passengers) in order to get somewhere in a hurry. This statement might seem callous to the true white scarfed airplane buff who flies for the pure enjoyment of flying, but nevertheless it is true. The percentage of the total population traveling by air has increased pretty much in proportion to the way that cruising speed has increased. Many airplanes are used for purposes other than rapid transit, but the fact remains that most of them are used to get from Point A to Point B; therefore, one of the first items of performance that the designer has to consider is its cruising speed.

As far as the airframe is concerned, in-flight performance depends on its drag characteristics. Total drag of an airplane is explained in chapter 3 and is graphically illustrated by the total drag curve in Fig. 3-13. The curve shows quite clearly that there is a speed associated with minimum drag (the bottom of this curve) and that drag increases with either an increase or decrease in speed from this point.

Chapter 4 said that the output of reciprocating engine-propeller powerplants is more comfortably handled in terms of power rather than thrust; thus, it is also convenient to represent the drag of the airplane in terms of the amount of power required rather than the amount of drag to be overcome. Power is the force required to move something a specific distance per unit of time (a specific number of feet per second, for example). Power can be considered as a force times velocity; therefore, the power required at any given velocity can be determined by multiplying the drag times the velocity. When this is done with a number of values of drag and corresponding velocity, and then plotted, the curve appears as shown in Fig. 5-1. This is the *power required curve* and is more commonly known simply as the famous *power curve*.

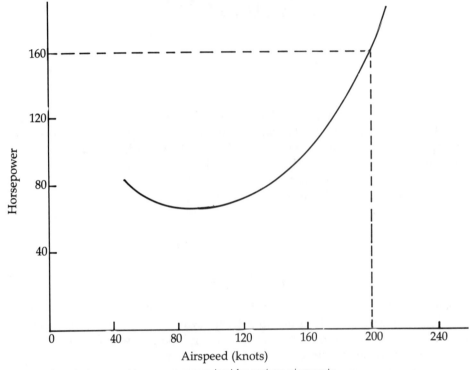

Fig. 5-1. A typical curve of horsepower required for various airspeeds.

The power required curve has nothing to do with engine power. The curve is simply the drag curve replotted in terms of power. Gliders have power required curves associated with them, just like airplanes, even though they have no powerplant. The power required curve is somewhat similar in shape to the drag curve, but skewed a bit due to the additional velocity term that it contains. Like the drag curve, the power curve has a minimum point. More power is required at speeds more than and less than the speed associated with minimum power; however, the speed at which minimum power is required is not the same as the speed for minimum drag.

The term power required actually refers to the power required for level flight; therefore, in order for the airplane to fly at a certain airspeed, there must be the corresponding amount of power available from the powerplant as indicated by the power required curve. Figure 5-1 shows a power required curve for an airplane that indicates to fly at 200 knots, a power of 160 horsepower is required. Obviously, if we only had a 100 horsepower engine, we could never attain this speed in level flight.

We could not reach this speed even if we had an engine rated at 160 hp. Remember that the rating is given in the brake horsepower (bhp) output of the engine. The actual power available for thrust is obtained by multiplying bhp times propeller efficiency. Because the peak efficiency of a propeller is approximately 80 percent, we would have to use a 200-bhp engine. Eighty percent of 200 is 160, so that is the maximum thrust horsepower (thp) that would be available.

Figure 5-2 illustrates this situation. The dotted line shows a constant value of 200 horsepower, which represents the maximum bhp of the 200 horsepower engine. Not all of this 200 horsepower is available to overcome the drag at any speed. The amount available is the thp, which is bhp times propeller efficiency. Propeller efficiency is always less than one, so the thp is always less than bhp. The efficiency is usually greatest in the cruising speed range and decreases at lower speeds. A curve can then be plotted for the power *available* and plotted on the same chart as the power *required*. Because the power available is the thrust horsepower available, the power available curve takes on the same shape as the propeller efficiency curve.

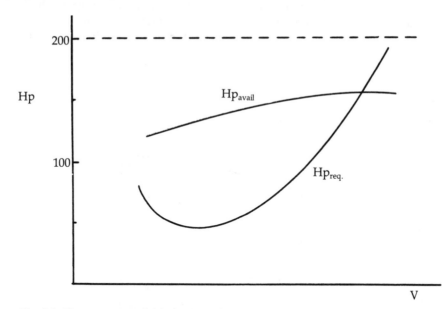

Fig. 5-2. Horsepower available for thrust for a 200 bhp rated engine shown along with horsepower required.

Note that the power available and power required curves cross. The speed corresponding to this point is then the *maximum* level flight speed. Above that speed, more power is required than is available, so it simply cannot be attained.

The power available curve in Fig. 5-2 represents the maximum power available at various speeds. In other words, it is the situation for full rpm and wide-open throttle. We don't normally cruise at this power setting. Recall from chapter 4 that the maximum recommended cruise power is usually approximately 75 percent of rated brake horsepower; therefore, the maximum power available for cruise from the 200-bhp engine is 75 percent of 200, or 150 hp. This figure is then reduced by the loss of the propeller.

With an assumed 80 percent efficiency, the thp available for cruise turns out to be a maximum of 120. Figure 5-3 shows that the value of 120 horsepower corresponds to about 180 knots; thus, while full bore would yield 200 knots maximum speed, a more respectable setting of 75 percent would yield a cruise speed of 180 knots.

Note that power could be reduced even further. The level flight speed attained would be whatever speed results where the power available curve intersects the power required. Level flight calls for the airplane to be somewhere on the power required curve. Sustaining the speed demands the required amount of power being delivered.

Note also that reduced airspeed requires less power only to the point of minimum power. This speed is reached at approximately 80 knots on the example in Fig. 5-3. Below that speed, increasingly more power is required to sustain level flight. This region of flight is known as the infamous "back side of the power curve," so named because the curve changes direction and curls up on the low-speed side of the speed range. This effect is due, of course, to increased induced drag in this region and results in a reversed trend in power requirement with airspeed.

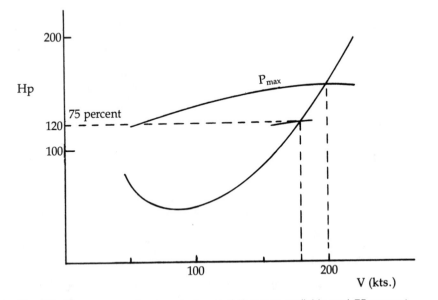

Fig. 5-3. Power curves showing maximum (full) power available and 75 percent power, and the resulting forward speeds.

The back side (*region of reversed power requirement*) is usually very short for many airplanes. This is because the minimum power speed is very low, and many airplanes stall before getting very much below that speed. Many pilots seldom get very far into the back side of the curve and do not experience reversed power demand to any degree. Changing to another aircraft, such as one with high induced drag, might present an entirely different situation and cause some surprises for unfamiliar pilots in slow flight maneuvers. Airplanes have many eccentricities that are not always obvious. This is one good reason why a thorough checkout should precede any transition to a new airplane, even one of seeming resemblance to your more familiar old model.

On the high-speed side of the power curve (the "front side," if you will), most of the drag is due to parasite drag, which is proportional to the square of velocity. Because power required is drag times velocity, the power then becomes proportional to an additional power of velocity or the cube.

$$D \propto V^2$$

$$P = D \times V \propto V^3$$

This fact indicates that considerable additional power is required to increase the top speed, or high cruise speed, of an airplane. A 33 percent increase in power results in only an approximate 10 percent increase in speed. Converting an existing airframe to a higher-powered engine for more speed is often a frustrating experience. The higher weight of higher-powered engines actually reduces the speed increase to even less. You could end up adding only a few knots of speed with an additional 100 or so horsepower.

# CLIMB PERFORMANCE

While speed is important, it is not the only item of performance to consider in evaluating an airplane. Too many pilots compare airplanes by looking only at the maximum cruise figure. Because flying involves getting up into the air, getting up there in a hurry would seem to be a great virtue of a true flying machine. Even if an airplane has high cruise speed, if it is very slow in reaching cruise altitude, much of the advantage of the high cruise is lost during the climb. High-altitude airports and airports boxed in by obstructions also require good climbing performance. The importance of such performance is almost equal to that of top speed.

The airplane's ability to climb is also determined by the power curves. It was pointed out that for level flight, just enough power must be available to match the amount of power required at any airspeed. Obtaining this correct amount of power is accomplished by adjusting the power controls; however, at less than maximum speed, more power is available than is required for level flight. The additional amount of power available is climb capability and the rate of climb is determined by the amount of additional power.

Figure 5-4 again shows the power required curve and the *maximum* power available curve, obtained from wide-open throttle and full rpm. With this power setting we could only fly level at top speed: the point where the curves cross. Anything less than that speed is *excess power*—more power available than required. Excess power that is not required for level flight is available for climbing. Excess power is sometimes termed *power differential* ($\Delta P$), the difference between maximum power available and power required. The rate of climb at any speed is proportional to this amount of excess power, $\Delta P$, and inversely proportional to weight.

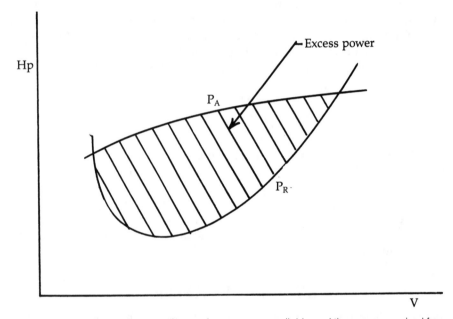

Fig. 5-4. Excess power between the maximum power available and the power required for level flight.

To show how rate of climb is determined, consider the example in Fig. 5-5. At 100 knots the power required for level flight is 40 horsepower, but the engine is capable of delivering 100 thrust horsepower at this speed; therefore, the excess power, $\Delta P$, is 60 horsepower. Rate of climb is simply excess power divided by weight. Horsepower divided by weight would actually give us units of horsepower per pound, not a very meaningful measure of rate of climb. To convert this figure into feet per minute, the scale that we are used to seeing on vertical speed indicators, multiply by 33,000. This number represents one horsepower converted to foot-pounds per minute.

Returning to our example situation, where we had a $\Delta P$ of 60 horsepower, and assuming that the airplane has a gross weight of 3,000 pounds, we can calculate the rate of climb. We divide the 60 excess horsepower by 3,000 pounds and multiply by 33,000 to yield a rate of climb of 660 feet per minute. Designating rate of climb as RC:

$$RC = \frac{\Delta P}{W} \times 33{,}000$$

$$= \frac{60}{3000} \times 33{,}000 = 660 \text{ ft/min}$$

This rate of climb would be possible at the selected airspeed of 100 knots; however, notice that the excess power is 60 horsepower only at this speed, and is not the same for other speeds. Because of the shape of the power curves, the excess power varies with speed. Different values of rate of climb would thus result at different airspeeds. Where the excess power is greatest, maximum rate of climb is available.

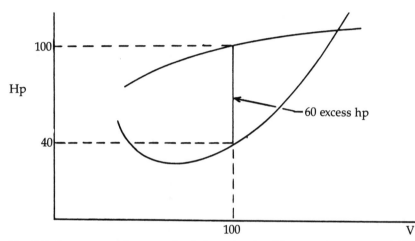

Fig. 5-5. Power curves for example airplane showing 60 excess horsepower at 100 knots airspeed.

Figure 5-6 depicts this situation. At one velocity we obtain the maximum rate of climb. The speed associated with this condition is called $V_y$ in aircraft performance specifications. This point occurs where the greatest $\Delta P$ exists, or where the power available and power required curves are farthest apart. Note that this speed is not necessarily the same as that for minimum power (the bottom of the power required curve). If we change speed in either direction from the $V_y$ point, the excess power becomes less and rate of climb becomes progressively lower.

Figure 5-7 shows a plot of rate of climb versus velocity as would be determined if we actually calculated it using various values of $\Delta P$ from a chart such as Fig. 5-6. The peak point on this curve is the maximum rate of climb and occurs at $V_y$. We can also determine from this plot the velocity for best angle of climb. *Best rate of climb* is the most altitude that can be obtained in a given amount of time. *Best angle of climb*, on the other hand, gives the most altitude for a given horizontal distance covered. This is a very handy speed to know if you have a large hickory tree at the end of a 2,000-foot strip. Best angle of climb airspeed is the recommended takeoff speed for short field performance when obstacle clearance is a factor.

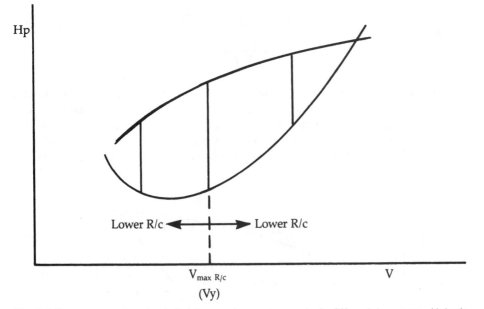

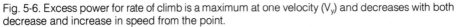

Fig. 5-6. Excess power for rate of climb is a maximum at one velocity ($V_y$) and decreases with both decrease and increase in speed from the point.

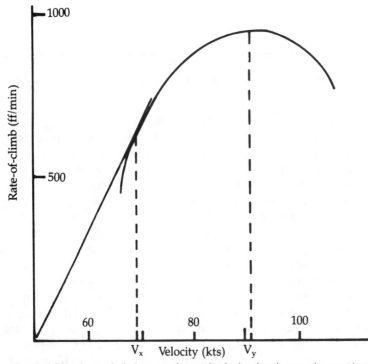

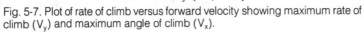

Fig. 5-7. Plot of rate of climb versus forward velocity showing maximum rate of climb ($V_y$) and maximum angle of climb ($V_x$).

Best angle of climb speed can be obtained from the curve in Fig. 5-7 by drawing a line from the point where RC and velocity are zero to the point on the curve where the line just touches it, or becomes tangent. At this point, the ratio of vertical speed, RC, to horizontal speed, V, is greatest. This means that the airplane operating here will climb the fastest in proportion to how fast it moves over the ground. Notice that this speed, $V_x$, is always less than the speed for best rate of climb, $V_y$.

Figure 5-8 shows two airplanes, each of which has lifted off the ground at the same point on the runway. The airplane on the right is climbing at best rate of climb speed. In a certain amount of time after takeoff it has reached the altitude and distance shown, following the path as indicated. The airplane on the left lifts off at the same spot but climbs at $V_x$, best angle of climb speed. In the same amount of time as the airplane on the right, this airplane has climbed to a lower altitude because its rate of climb is less, and has covered less ground because its forward speed is less; however, the path of flight makes a steeper angle with the ground. At any given distance beyond the takeoff point, its altitude will be higher than the airplane at $V_y$.

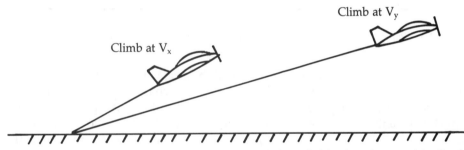

Climb at $V_y$

Climb at $V_x$

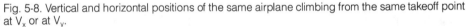

Fig. 5-8. Vertical and horizontal positions of the same airplane climbing from the same takeoff point at $V_x$ or at $V_y$.

Climbing at $V_x$ is normally only done for a short period of time immediately after takeoff, if obstacle clearance is a problem. For most climbing operations, rate of climb is the major consideration. The power curves that have been discussed so far, from which we determined rate of climb, are really representative of one altitude. Let us assume that they represent sea level performance. The rate of climb of 660 ft/min that was calculated from the 60 excess horsepower would then be the value at sea level (for the velocity and weight specified). As we go up in altitude, power required and power available changes. The power available decreases for all airspeeds. Power required becomes greater at low speeds and less at high speeds, again because of the relative effect of density on induced and parasite drag.

The change in power curves for higher altitude is shown in Fig. 5-9. The basic effect is that the power curves come closer together, resulting in less excess horsepower. The maximum $\Delta P$ at altitude is obviously much less than that at sea level. The effect on rate of climb is a similar reduction because rate of climb is directly proportional to excess power.

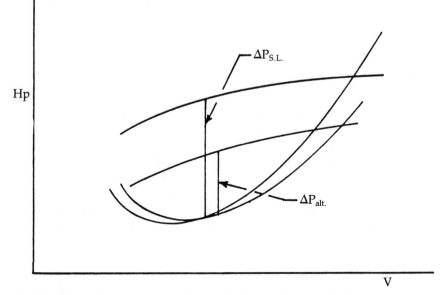

Fig. 5-9. Change in power curves at higher altitude showing reduction in excess power.

Notice also that the speed associated with best rate of climb, $V_y$, is increased slightly with altitude. The velocity on any of these charts is implied to be *true* airspeed. $V_y$ in true airspeed does increase with altitude. Because true airspeed is related to calibrated airspeed through the density ratio (the ratio of density at altitude to that at standard sea level), variation of $V_y$ in *calibrated* airspeed with altitude is not as great. In many airplanes, $V_y$ in calibrated airspeed is almost constant with altitude. Because this is essentially the value read on the indicator (corrected for minor instrument error), nature, for a change, makes it easy on the pilot and requires him to remember only one value of $V_y$.

Returning to our discussion of altitude effect on rate of climb, let us consider proceeding to an even higher altitude. The power curves now become even closer, as shown in Fig. 5-10. Excess power is reduced even more, with resulting decreased rate of climb. Another consequence also becomes evident in this illustration. The range of airspeeds where there is excess power—and hence, climb capability—is becoming less. This range is indicated from $V_a$ to $V_b$ in Fig. 5-10. The necessity to maintain a speed of $V_y$, or close to it, is becoming more critical.

Finally, if climb is continued, an altitude is reached where the curves just barely touch, as depicted in Fig. 5-11. At this point there is no more excess power and climbing ability is zero. The altitude where this occurs is called the *ceiling*. The point where the curves touch is the required level flight speed, and even this is only possible if the airplane is right at this very critical speed. Obviously, flight at this extreme limit is not very practical, even if it could be maintained by a skillful pilot. This limiting altitude is the *absolute* ceiling and is not considered within the realm of operation.

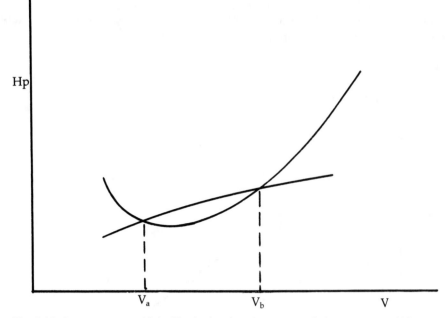

Fig. 5-10. Power curves at high altitude showing short range of airspeeds ($V_a$ to $V_b$) over which excess power exists.

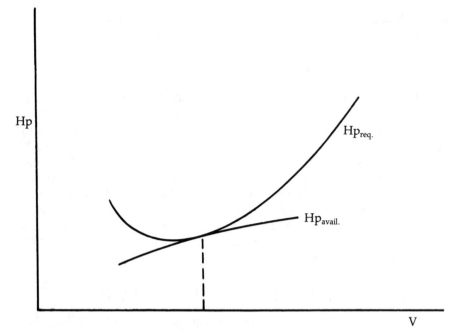

Fig 5-11. At the absolute ceiling no excess power exists and only one airspeed will maintain level flight.

A more practical limit on altitude is the *service ceiling*, which is the altitude where the rate of climb is only 100 feet per minute. The variation in maximum rate of climb with altitude for a typical lightplane is displayed in Fig. 5-12, with absolute and service ceilings indicated.

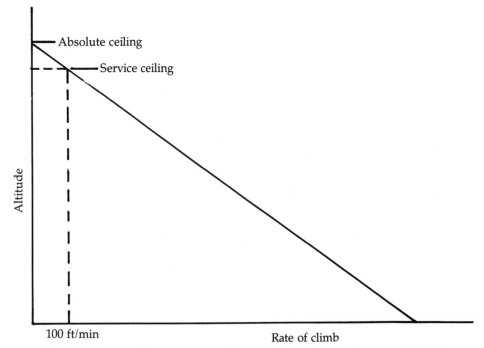

Fig. 5-12. Variation of rate of climb with altitude. Absolute ceiling occurs where rate of climb is zero and service ceiling occurs where rate of climb is 100 feet per minute.

All of the power required curves shown have considered a constant value of gross weight. Most published airplane performance is for maximum (or design) gross weight. Lower gross weight will also have an effect on climbing ability because it also changes the shape of the power required curve. Lower weight reduces induced drag, which is most significant at low speeds, but has some effect over the entire speed range. The effect of reduced weight on the power required curve is an overall reduction, but greatest at low speeds, as shown in Fig. 5-13. Lower power required leads to greater excess power and results in higher rate of climb. This effect could also be deduced from the formula given for rate of climb. Weight is in the denominator, which, if reduced, would inversely affect RC, or cause it to increase.

## TWIN-ENGINE CLIMB PERFORMANCE

A misconception among the uninitiated—even among some pilots—is that additional engines are added to airplanes to provide some safety factor, or a degree of redundancy. While this factor might be of some consideration, the primary rea-

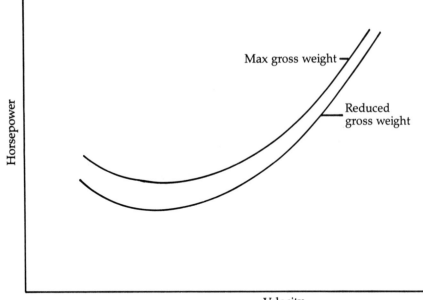

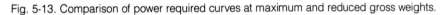

Fig. 5-13. Comparison of power required curves at maximum and reduced gross weights.

son for multiengine configuration is simply to provide more power, which is required by larger airplanes. Engines can only be made so large and still operate efficiently. When more power is required than can be accommodated in one engine, an additional engine is added and the design becomes a twin.

Power requirements are just as they are with any airplane. The only difference is that the power available is provided by two smaller engines rather than one big one. Much of the power of the two engines is required to maintain level flight, just as it is with single-engine configurations. Excess power above this amount contributes to climb performance.

Figure 5-14 shows a typical power required curve and the maximum power available with both engines. The rate of climb at a given velocity is proportional to the excess power available from both engines, indicated as $\Delta P(2)$. Now suppose one engine fails. Many who have not handled twin fans assume that the rate of climb would be cut in half. This assumption is quite erroneous. A loss of one engine results in a loss of half of the power available; however, much of that power available was required just for level flight, shown as $hp_{req}$. Now all of this power has to be supplied by one engine. The remaining power available from the single engine $hp_A$ (1 eng) is quite often not very much more than that required for level flight. The result is a much reduced amount of excess power, $\Delta P(1)$, and consequently, a much reduced climb capability.

Many light twin aircraft suffer a reduction in rate of climb of approximately 70 percent when one engine fails, some as much as 90 percent. Ceiling suffers also, and is reduced to a few thousand feet for some twins. In addition to the loss of power, the trim required to correct for the asymmetric thrust also adds to drag,

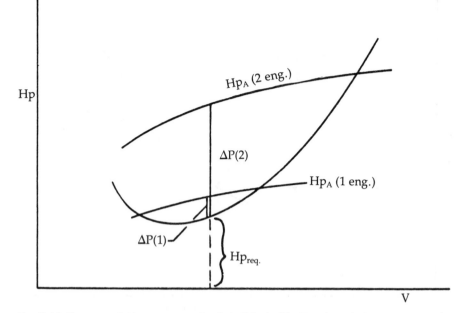

Fig. 5-14. Power available on one engine is half that of two engines, but excess power is much less due to the same amount of power required for level flight.

making climb performance even worse. Twin-engine aircraft can be flown on one engine, and while this does add a bit of safety factor, the performance is severely limited, reducing the safety factor to a rather slight one. *That additional engine is not meant to be a spare.*

## DESCENT AND GLIDE PERFORMANCE

So far we have talked about excess power, or the case where power available is greater than the power required. Let us now see what happens when power available is less than required. In Fig. 5-10 we saw that climb was possible only between the speeds indicated as $V_a$ and $V_b$. Precisely at these speeds, level flight could be sustained. Incidentally, it should be noted that there are conditions (such as shown here) where two different level flight speeds could be maintained with the same power setting. This situation would be more likely encountered at high altitude, where high induced drag causes a high back side of the power required curve.

When flying faster than $V_b$, or slower than $V_a$, power available is less than power required. In this case we have a decrement of power, rather than excess power. There still exists a difference in power, $\Delta P$, but it is now negative. We could still calculate rate of climb as $\Delta P$ divided by weight, but with a negative $\Delta P$, we get a negative rate of climb. With negative rate of climb we are not climbing at all, we are sinking, and the resulting vertical velocity is called *rate of sink*. In this case:

$$RC = \frac{-\Delta P}{W} \times 33{,}000 = RS$$

Because the same formula is involved, the rate of sink is again proportional to the $\Delta P$, or the difference between power required and power available. Notice that the further the speed gets from $V_a$ or $V_b$, the greater this difference becomes.

Rate of sink could be affected even between the speeds of $V_a$ and $V_b$. If, instead of changing speed, the power were reduced to make the power available less than power required, rate of sink could be established at any speed. Power can be reduced all the way to zero. This is the situation with power reduced to idle—always the case for a glider.

With power available reduced to zero, the power available curve is actually the horizontal axis of the graph as shown in Fig. 5-15. The power required curve does not change, so $\Delta P$ is now the distance between zero and the power required at any velocity, meaning simply the value of power required. Because rate of sink is proportional to the $\Delta P$, it is obviously lowest where the power required is lowest. This occurs at the bottom of the curve, and the velocity at this point thus gives us the minimum rate of sink. The actual minimum rate of sink is proportional to the $\Delta P_{min}$ at this point, as shown.

Minimum rate of sink does not necessarily mean the best overall glide performance. Because it is a very low airspeed, the aircraft is moving forward very

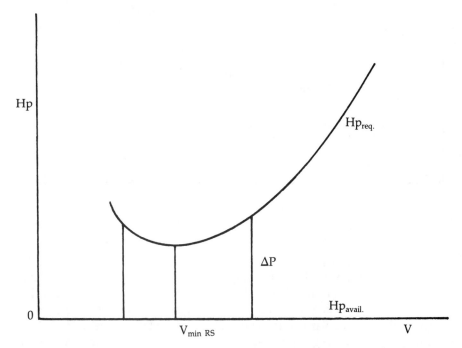

Fig. 5-15. Power required and available for glider or power-off situation, showing decrement in power ($\Delta P$).

slowly at minimum sink speed. The distance covered during the glide could be increased by increasing velocity slightly. The maximum glide distance would actually be obtained at the minimum ratio of rate of sink to forward speed. This point can be located somewhat similar to the method of finding best angle of climb. A straight line is drawn from the point where both P and V are zero (the origin of the axes of the graph) to where it just touches the power curve (becomes tangent), as shown in Fig. 5-16. The speed corresponding to this point is the *best glide speed*, or speed for maximum glide distance.

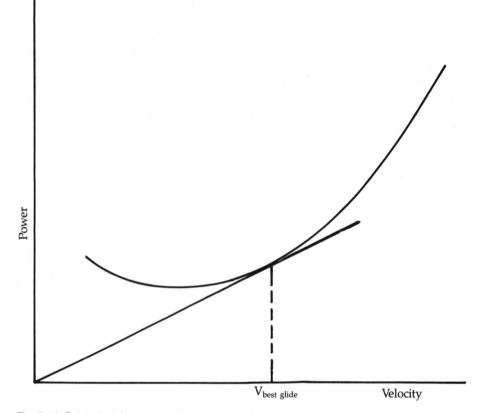

Fig. 5-16. Point of minimum ratio of power to velocity on the power curve determines the speed for maximum glide distance.

The power required at this point is the $\Delta P$ and determines the rate of sink for maximum glide distance. Flying at this point gives you the greatest horizontal distance for a given amount of altitude. This ratio of horizontal to vertical distance is known as *glide ratio*. This aspect of gliding performance is usually of most concern to pilots, particularly in the event of an engine failure over a mountain or the middle of town. Flying at the speed for minimum sink rate, on the other hand, would give the maximum time that you could remain airborne in a glide. All of this discussion, of course, concerns true gliding in static air. Soaring in vertical currents, as practiced by sailplane pilots, adds a whole new dimension to powerless flight. Such performance is beyond the scope of this book, although the basic principles still apply.

Another fact about gliding performance is significant. We discussed the fact that the minimum ratio of power to velocity gave us the greatest gliding distance or glide ratio. Power required is equal to drag multiplied by velocity, so that a ratio of power to velocity is the same as drag times velocity divided by velocity, which results in, simply, drag.

$$\frac{P}{V} = \frac{D \times V}{V} = D$$

Thus, the point of minimum power to velocity ratio is also the point of minimum drag; the ratio of power to velocity, $P/V$, at any velocity is the drag at that speed. Obviously, the minimum $P/V$ is the same as minimum drag. Maximum glide ratio occurs, then, at the speed for minimum drag, which seems to make sense.

Another way of looking at the minimum drag speed is to consider the lift to drag ratio, $L/D$. Lift in level flight is equal to the weight. Even in the normal angles of glide, lift is almost equal to weight. Only at high dive angles does it vary significantly; therefore, throughout the entire range of possible flight speeds, the lift is essentially constant at a particular gross weight. The $L/D$ ratio would thus vary only with change in drag. At minimum drag, $L/D$ would be a maximum; therefore, maximum glide ratio also can be thought of as occurring at the velocity for maximum $L/D$ ratio. Furthermore, it works out mathematically that the $L/D$ is exactly equal to the ratio of forward velocity to vertical velocity (rate of sink). The ratio of $V$ to RS is also equal to the ratio of horizontal distance to altitude (glide ratio):

$$\frac{L}{D} = \frac{V}{RS} = \frac{\text{glide distance}}{\text{altitude}}$$

Knowing any three of these numbers would enable you to compute the fourth one, as in this example:

$$\text{glide distance} = \text{altitude} \times \frac{L}{D}$$

When working with ratios, though, the terms must be in consistent units. That means vertical and forward velocities must be in feet per second, or miles per hour, or other unit of velocity.

# RANGE

An airplane that climbs and cruises well is of little value if it can only go 100 miles or so before it must be refueled. Range, then, is another consideration in an efficient airplane design. Range cannot be extended simply by adding more fuel capacity without affecting other performance. Fuel adds weight to the aircraft and weight deteriorates most performance. The desired range must be considered by the engineer in the initial stages of design. Range can be varied, however, for a given fuel capacity. Operating at certain combinations of power and velocity can increase range over other, less efficient settings.

Chapter 4 points out that an engine has a specific fuel consumption, sfc, of a certain number of pounds of fuel required per brake horsepower developed per hour.

$$SFC = \text{lbs of fuel/bhp/hr}$$

The actual fuel consumption per hour, FC, is the actual bhp being developed times the specific fuel consumption.

$$\text{fuel consumption} = FC = SFC \times bhp = \text{lbs of fuel/hr}$$

If we divide the fuel consumption into the cruising speed, we would get the number of miles flown per pound of fuel consumed. This term is known as specific range, SR.

$$\text{Specific range} = SR = \frac{V}{SFC \times bhp} = \text{miles/lb}$$

The total range is then simply the amount of fuel on board (in pounds) multiplied by the SR.

$$\text{Range} = R = \text{lbs of fuel} \times SR = \text{miles}$$

This last calculation is fairly straightforward; if fuel tanks are always filled completely, the fuel quantity used in this formula would always be the same; however, specific range will vary, depending on cruising conditions.

Note that the equation for specific range includes the ratio of velocity to power (V/bhp). Power was referred to simply as P in the section on glide performance. There it was actually implied to mean thrust horsepower, which differs from bhp only by the propeller efficiency. At any rate, it was shown that the ratio of P/V was actually another way of expressing drag. The inverse of this term, V/P, multiplied by weight becomes the L/D ratio. For a given weight, greater V/P ratio means greater L/D ratio. According to the formula above, the specific range increases with V/bhp, so it also depends on L/D ratio. The maximum specific range would thus be made possible at the maximum L/D ratio.

The maximum L/D point on the power curve for cruise flight is found in just the same way as it is for a glider because the power required curve is essentially the same. Figure 5-16 shows a line from the graph origin point to a tangency point on the power curve to determine best glide speed.

This speed yields maximum L/D. We can conclude that operating at maximum L/D gives best glide and best range. Basically, maximum L/D speed gives the greatest possible flying distance with power on or off.

For range considerations, we are concerned with the power-on situation. In order to cruise at a certain velocity, we must set the power available to that required at that velocity; hence, to fly at the speed for best L/D, we must also cruise at the power for best L/D. Unfortunately, the speed and power setting for maximum L/D are far below normal cruise values for most airplanes. As a result, very, very few aircraft ever actually cruise at this condition and thereby achieve maximum range.

Figure 5-17 shows that the maximum range speed for the Piper Arrow II is 97 knots. The power required is also 76 thp, which converts to approximately 100 bhp, or 50 percent power. At 75 percent power this airplane will cruise at approximately 133 knots. Obviously, operating at maximum range speed is a pretty slow way to travel. It is only practiced when long stretches of unlandable areas are crossed, such as open water, or by those of the penny-pinching persuasion.

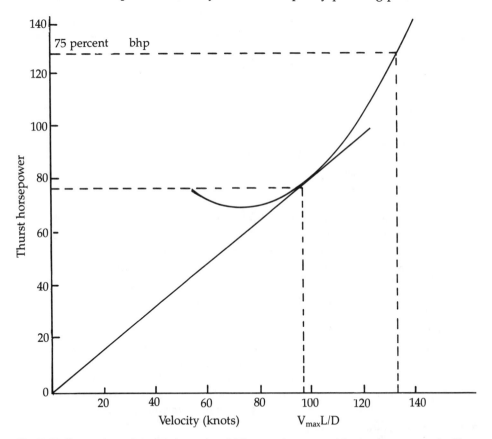

Fig. 5-17. Comparison of speeds for cruise at 75 percent power and for maximum range for Piper Arrow II , as determined from power curve.

Another problem arises in attempting to cruise at maximum L/D. The power curve from which a maximum L/D is determined is for a given value of weight. As fuel is consumed in flight, weight decreases. On a long-range flight, the weight change can be considerable; therefore, the power curve is different at the end of the flight and the resulting velocity for maximum L/D actually is less than at the beginning of the flight. Figure 5-18 shows a typical decrease in maximum L/D speed from takeoff at maximum gross weight to landing with most fuel consumed.

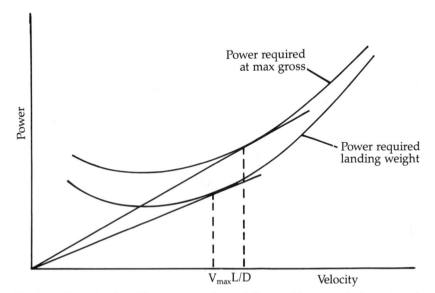

Fig. 5-18. Chart showing difference in maximum L/D speed for same airplane at maximum gross weight and at landing weight with fuel consumed.

With long-range aircraft, where fuel weight is a high percentage of gross weight, this continually changing speed and power setting for best L/D can have a significant effect on range. Many light airplanes, however, have a fuel capacity of only approximately 10 to 15 percent of the gross weight. In this case, the velocity for best L/D changes very little with fuel burnoff. The Piper Arrow cited above, for example, has a decrease in speed for best L/D of only 2 knots when the weight of the fuel is subtracted from full gross weight.

Another factor that affects maximum range is wind. A headwind or tailwind will change the speed for best L/D. To take an extreme case, consider a headwind of the same intensity as the best range speed for a no-wind condition. Flying at this speed would now give us a range of zero. Obviously, it is no longer the best range speed. In order to overcome the headwind, a speed somewhat greater would have to be flown. The best speed could be determined in this case if a line to the tangent point is started, not from the origin, but from a point corresponding to the value of the headwind on the velocity axis of the graph. This would result in a higher speed for best L/D.

Conversely, a tailwind would shift the start of the line a negative distance from zero velocity equal to the value of the tailwind. The resulting line would touch the curve at a point of lower velocity than for a no-wind case. Figure 5-19 illustrates this procedure and resulting velocities. Remember that this only applies to operation at maximum range (best L/D) speed and power. It does not imply that cruise speed should be reduced with tailwind and increased with headwind when operating at higher power settings.

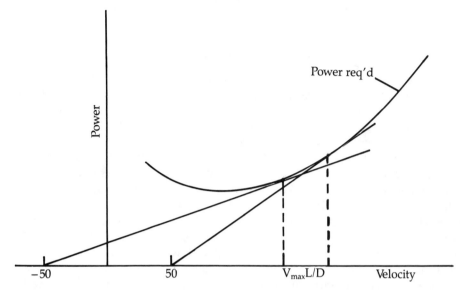

Fig. 5-19. Chart showing difference in maximum L/D speeds for airplane with 50 knot headwind and 50 knot tailwind.

## ENDURANCE

Range is the *distance* that an aircraft can fly on a given amount of fuel. The *time* that an airplane can remain aloft on a certain amount of fuel is given the term *endurance*. If the range is known for any particular velocity, then the endurance could be determined simply by dividing the range by the velocity.

$$\text{Endurance} = \frac{R}{V}$$

The *maximum* possible endurance, however, is not obtained at the speed for maximum range; therefore, maximum endurance is not obtained by using maximum range and best L/D speed in the above formula. Fuel consumption is proportional to power in a propeller-driven airplane. The minimum amount of fuel would thus be obtained at minimum power. Minimum fuel consumed per hour would give the most hours of flight; therefore, maximum endurance is obtained when operating at minimum power required. This point is the very bottom of the power required curve. Referring again to Fig. 5-17, the Piper Arrow appears to have a maximum endurance speed of approximately 74 knots at 68 thrust horsepower. This speed is considerably slower than that for maximum range.

If the endurance is desired and the range is not known, it can be calculated directly as the fuel amount available divided by the fuel consumption. Fuel consumption at a given power setting is SFC times the bhp:

$$\text{Endurance} = \frac{\text{fuel amount}}{\text{lbs/hr}} = \frac{\text{fuel amount}}{\text{SFC} \times \text{bhp}}$$

## CRUISE EFFICIENCY

As explained above, cruising at maximum range or maximum endurance speed is an extremely slow progress. Except for emergencies or unusual circumstances, it is hardly ever practiced. Even if it were attempted, determining the exact speed and power settings for the appropriate weight and altitude throughout the flight is an involved process. Most flying is done at a certain power setting chosen for the desired performance. Higher power settings yield higher cruise speed, but greater fuel consumption and less range. Lower settings give better economy and longer range, but reduced speed. There are ways to compromise, however, and obtain the best combination of speed and economy.

Fuel consumption is proportional to brake horsepower. The gallons of fuel—and corresponding dollars per gallon—that burn each hour will depend upon the percent of brake horsepower being developed. Seventy-five percent power will consume 20 percent more fuel than 55 percent. The speed—and consequently, distance covered—depends not only on power setting, but also on the altitude. Higher altitude reduces drag, and corresponding power required, at speeds in the range of normal cruise. Higher altitude also reduces the maximum power that the engine can develop; however, 75 percent power is 75 percent of maximum brake horsepower. As long as you can still get that much out of the engine, the higher the aircraft, the faster it will cruise at that power setting.

Figure 5-20 shows a typical variation of airspeed with altitude for various power settings. The curved line represents the maximum speed available with full power. It is a maximum (140 kts) at sea level and decreases with altitude; however, cruising at more than 75 percent power is not recommended for this airplane (as is the case with most piston engines), so that 124 knots is all that could be obtained at that setting at sea level. With increased altitude, however, the speed at 75 percent power increases. At 7,500 feet, the 75 percent line and the maximum speed curve intersect, indicating that this is the limit for 75 percent power. Above 7,500 feet the engine will not develop 75 percent. At this intersection, the airspeed is 133 knots. No change in fuel consumption would be experienced at the constant power setting of 75 percent. Going from sea level up to 7,500 feet, though, would increase cruising speed from 124 to 133, a gain of approximately 7 percent. Cruising at this altitude would reduce the flight time, and, at the same power and fuel consumption, would save fuel.

Lower power settings may be maintained to higher altitudes. Sixty-five percent power can be maintained up to approximately 10,000 feet and 55 percent will hold out until almost 12,000. At 12,000 feet, cruising at 55 percent will result in 118 knots, but consume no more fuel than cruising at sea level, where the speed would only be 105. Clearly, the highest possible altitude for a given power setting

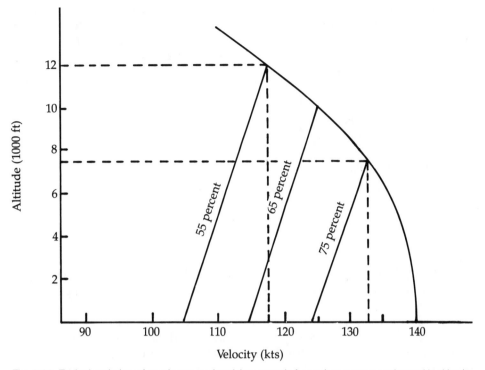

Fig. 5-20. Typical variation of maximum and cruising speeds for various power settings with altitude.

is the most economical. Lack of oxygen or air traffic control restrictions and assignments might not always make the exact optimum altitude available; however, getting as close as possible would maximize the economy for a particular flight. It is important for the pilot to become familiar with the performance charts in the flight handbook so that these altitudes are known.

Another consideration of fuel-efficient cruising is due to some recent work of Dr. Bernard H. Carson, who teaches at the U.S. Naval Academy. Carson recognized the very slow speed associated with maximum L/D and sought to find a faster speed that would yield some optimum combination of speed and economy. He reasoned that operating at maximum L/D produced the minimum fuel consumption per *mile* traveled; however, the minimum fuel consumption per *unit of speed* (say, per knot) would result when operating at the maximum ratio of velocity to drag, V/D. Maximum V/D ratio can be obtained from the plot of drag versus velocity. A line is drawn from the origin to the tangency point of the drag curve similar to the way that maximum L/D velocity is obtained from the power curve; however, using the drag curve, the resulting velocity is that for minimum D/V, which is the same as maximum V/D. This velocity was termed the *cruise-optimum airspeed*.

Cruise-optimum airspeed gives the best return in increase in airspeed for excess fuel consumption above the optimum. Carson referred to operation at this speed as "the least wasteful way of wasting fuel."

It turns out from mathematical relationships that this cruise-optimum airspeed is 1.316 times (or roughly $1/3$ greater than) the velocity for maximum L/D, or best range speed. Best endurance speed is, interestingly, maximum L/D speed *divided* by 1.316; therefore, it appears that three velocities are of importance in most efficient flight conditions:

- Best endurance speed, $V_E$
- Best range speed, $V_R$
- Cruise-optimum speed, $V_C$.

These speeds are related:

$$V_E = \frac{V_R}{1.316}$$

$$V_C = 1.316 \times V_R$$

Operating at $V_E$ consumes the least amount of fuel per unit of *time*; $V_R$ yields the least fuel consumption per unit of *distance*, and $V_C$ yields the least fuel per unit of *velocity*.

Interestingly enough, $V_C$, the cruise-optimum speed, turns out to be just about the speed for 75 percent power at any altitude and full gross weight for most light airplanes manufactured in the 1980s. At less than full gross weight, the cruise-optimum airspeed decreases. At approximately the minimum operating gross weight, $V_C$ is very close to the speed for 55 percent power; thus, the normal operating range of speed and power from most modern airplanes actually is relatively efficient.

## TAKEOFF

Up to this point in this chapter we have been discussing *flight* performance. In-flight performance was shown to depend a great deal on the shape of the power required curve at the particular operating condition. Now let us consider a different type of performance. When the airplane is taxiing or beginning its takeoff, it is a ground-based vehicle. At the end of the takeoff it is a flight vehicle; therefore, takeoff is the transition from ground operation to flight and becomes a special performance consideration. The principles of steady flight performance no longer apply.

In takeoff operation, the designer is concerned with making the transition to flight as short as possible; hence, takeoff *distance* is the primary consideration. An airplane will not fly below its stall speed. Obviously, then, the airplane must be at or above stall speed in order to take off. At the beginning of takeoff, the airspeed is zero (for a no-wind situation). The situation in taking off thus becomes one of accelerating from zero airspeed up to, or above, stall speed. The distance required to make this change in velocity is the *takeoff distance*.

During takeoff, the airplane has the usual four forces associated with it, as shown in Fig. 5-21; however, the forces are *not* in balance. In order for the airplane to accelerate, thrust must be *greater* than drag. The lift is also not equal to the weight below takeoff speed. Also, there is an additional force present during the takeoff roll, namely, the frictional force between the tires and the runway. The acceleration resulting in the takeoff roll depends on the relation in Newton's second law, which says that force equals mass times acceleration.

$$F = m \times a$$

Rearranging this equation, we can then deduce that acceleration equals force divided by mass. The force here is the *net* force in the direction of the acceleration, and this *net* force is equal to thrust minus drag and frictional force.

$$a = \frac{F}{m} = \frac{T - D - F_{frict}}{m}$$

Mass is simply weight divided by g, the gravitational acceleration, which at or near the Earth's surface is 32.2 feet/second; therefore, the acceleration can be written, in terms of weight, as:

$$a = \frac{g \times (T - D - F_{frict})}{W}$$

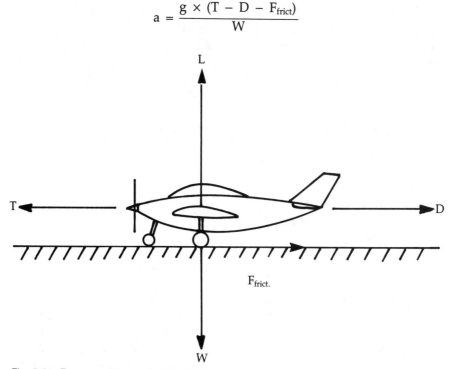

Fig. 5-21. Forces acting on airplane during takeoff.

Higher acceleration means a faster transition to flying speed and consequently shorter takeoff distance. The above equation shows that the acceleration term will increase with increased thrust and decrease with increased drag and frictional force because these last two forces subtract from the acceleration. Weight, being in the denominator, will decrease acceleration as it increases. The frictional force is also proportional to weight initially, although it is decreased as lift reduces the net weight during takeoff roll.

Thrust is therefore the only positive accelerating force; all the others tend to retard takeoff and increase the distance required. There is another consideration that must be made in takeoff; acceleration is required to reach (or slightly exceed) stall speed. If the stall speed could be reduced, it could be achieved more quickly and the takeoff distance reduced in this manner; therefore, with the same thrust, drag, and weight, takeoff could be made in a shorter distance if the stall speed were less. Stall speed is proportional to the square root of weight, but also inversely proportional to wing area and the maximum lift coefficient.

$$V_s \propto \sqrt{\frac{W}{S \times C_{L_{max}}}}$$

Therefore, increased wing area, S, or increased maximum lift coefficient will reduce stall speed and shorten takeoff distance. High maximum lift coefficient, remember, can be achieved with high-cambered airfoils or by use of flaps; however, high camber, either from flaps or from the natural airfoil shape, also increases drag, and drag detracts from takeoff performance. Usually, a small amount of flap deflection will increase lift (thereby reducing stall speed) with a relatively small increase in drag. Large flap deflections are primarily associated with drag and little additional lift. For this reason, a small amount of flap deflection can usually shorten takeoff distance, but greater than the optimum amount of flap deflection will increase takeoff distance.

To summarize, the following characteristics of the airplane will favorably affect takeoff distance:

- High thrust
- Low drag
- Low runway-tire friction
- Low weight
- High wing area
- High lift coefficient

In addition, runway characteristics and atmospheric conditions can also be quite significant factors in takeoff. Not all runways have the same surface conditions, making the frictional effect different. The romantic and nostalgic effects of a grass field takeoff can sometimes be outweighed by the increased friction of the grass, especially if the grass is tall. Runway slope also has an effect on takeoff. An uphill takeoff involves a component of weight, which also adds to the retarding forces and must be overcome by the thrust, thereby decreasing the acceleration. This additional force is equal to the weight times the sine of the runway slope angle. Of course, a decrease in retarding force of the same proportion occurs with a downhill run.

Thrust is greatest when air density is highest. Although higher density increases drag, the increase in thrust is greater; also, the stall speed is decreased. Therefore, high density, which results from high pressure and low temperature, works favorably in reducing takeoff distance.

Wind is also a deciding factor in takeoff. A headwind has the effect of reducing the takeoff distance by the amount normally required to accelerate from zero velocity up to the wind velocity. The airplane starts out with an airspeed already equal to the speed of the headwind component. To consider the extreme (but possible) case of a headwind equal to takeoff velocity, the takeoff distance would be zero. Under more normal conditions, the reduction in takeoff distance with headwind is given by the following formula:

$$\text{Takeoff distance} = \text{distance for zero wind} \times \left[\frac{V_o - V_w}{V_o}\right]^2$$

In this equation:
$V_o$ takeoff velocity with zero wind
$V_w$ velocity of headwind

Again summarizing, the environmental factors that favorably affect takeoff distance are as follows:

- Smooth runway surface
- Level or downhill slope
- High pressure
- Low temperature
- High headwind

Takeoff distance therefore depends upon quite a number of variables. Performance data provided pilots must take into account all of the above factors, plus the airplane factors listed previously.

The possible velocity for takeoff has been mentioned as being at or slightly faster than stall speed. Actually, no aircraft can lift off precisely at stall speed. Lifting off just a hair above stall speed can also be a somewhat frustrating experience. The airplane would settle back onto the runway at the slightest reduction in speed, and the resulting takeoff run would probably end up longer than normal. In order to ensure a positive liftoff, takeoff speed is usually considered to be 20 percent above stall speed. This, of course, has to be based on the stall speed in the takeoff configuration, including the particular flap setting involved. Takeoff data in flight handbooks are usually determined in this manner.

Another way that takeoff performance is sometimes given is the distance to clear a 50-foot obstacle. This distance includes the distance to break ground and the horizontal distance from there to where the airplane has gained 50 feet of altitude. The second part of this distance is determined by the rate of climb of the aircraft at the liftoff speed.

## BALANCED FIELD LENGTH

If an engine should fail during takeoff, the pilot of a single-engine aircraft has no alternative but to jam on the brakes and pray that there is still enough runway between him and the fence to get stopped. The pilot of a multiengine airplane has more options. If an engine fails fairly far down the runway, he might have sufficient speed at that point to safely continue the takeoff with one engine out. On the other hand, if the engine failure occurred early in the takeoff, there would probably be plenty of runway to abort the takeoff and come to a stop, however, in the middle portion of the takeoff, the decision of whether to continue with a failed engine or to abort becomes more difficult.

In order to provide multiengine pilots with a definite criterion on which to base this decision, a *critical engine-failure speed* is established, called $V_1$. Slower than this speed, the pilot aborts the takeoff and brings the airplane to a stop in the event of an engine failure. If an engine fails when faster than $V_1$, the pilot must continue the takeoff on the remaining engine(s). $V_1$ is established so that if the engine should fail precisely at that point, the distance required to stop is exactly the same as that required to reach takeoff speed, known as $V_2$.

The overall runway distance up to this point is known as the *balanced field length*. This distance is shown on the diagram in Fig. 5-22. The solid line shows normal acceleration with all engines operating, reaching $V_2$ well before balanced field length. The dotted lines represent the two possible profiles of velocity after engine failure at $V_1$. One is a reduced acceleration to $V_2$; the other a deceleration to zero velocity. Both of these results occur at the point of balanced field length.

Jet aircraft (actually, all aircraft certificated under FAR Part 25) are required to have a takeoff distance specified as the longer of either the engine-out accelerate-go distance or the decelerate-stop distance. Balanced field length, where both of these distances are equal, ensures the shortest field length requirement.

## LANDING

Landing performance is basically takeoff in reverse. It is a transition from flight to ground operation. The objective here is to decelerate from a velocity somewhat above stall to zero in as short a distance as possible.

The exact distance required for landing depends a great deal on pilot technique. One pilot might make the first turnoff, while another pilot flying the same aircraft under the same conditions might use most of the runway. Approach and touchdown speed, as well as the degree and rapidity of braking, make the difference. In order to specify landing distance in performance handbooks, certain speeds have to be assumed so that a possible landing distance can be established that can be accomplished with reasonable skill.

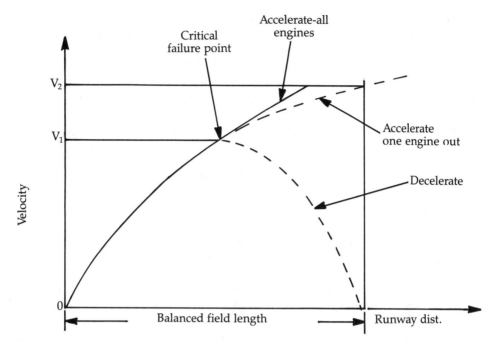

Fig. 5-22. Balanced field length is determined by establishing a critical engine failure speed so that the runway distance from that point to continue takeoff on the remaining engines or to decelerate and stop is the same.

In the good old days (or bad old days, depending on how you regard them) pilots were taught to make full-stall landings. Touching down at stall speed assured the shortest possible landing roll; however, this procedure required the final stage of the approach to be very close to stall. Operating close to stall speed on short final is not too safe and also makes handling rather difficult. Power approaches were standard for these airplanes and landing was assumed to occur at approximately 20 percent above stall speed, the same as takeoff speed. The final approach was also considered to be made at this same speed. Later, as performance increased even more, 30 percent above stall speed became the accepted (and required, in the case of transport aircraft) approach speed.

Landing distance, like takeoff, is also considered in two parts. The actual landing roll distance is the distance from touchdown to stop. Distance to clear a 50-foot obstacle is determined by adding the landing roll distance and the distance to descend from 50 feet of altitude. For light aircraft, a deceleration is usually considered to take place during this final stage of descent, reducing speed from 30 percent above stall at the 50-foot point to 15 percent above stall at touchdown. This method of approach shortens the distance of the airborne portion of landing. Touchdown at 15 percent above stall is more appropriate for modern aircraft. Pilots are usually taught this method of landing even with light trainers, so that the habit will carry over to later activities with heavier aircraft.

Once on the ground, the same forces are acting as we showed for takeoff in Fig. 5-2; however, the thrust will be reduced to zero and drag and friction will be the remaining forces in the plane of the motion. Because both of these factors are negative, the equation for acceleration turns out to be negative, or a *deceleration*. As the speed decreases throughout the landing roll, drag becomes increasingly low. Frictional force is the remaining factor that can save us from the fence at the end of the field. Normal rolling friction from the tires on the runway is usually not enough. Indeed, with a slight downward slope in the runway, the airplane could roll through eternity without some barrier to stop the aircraft.

Brakes are therefore a requirement for most landings and the rolling distance depends a great deal on the braking power available. Frictional force is usually determined by a coefficient of friction multiplied times the force perpendicular to the plane of motion. Again, as with typical coefficients, the friction coefficient represents how much of the force that the tire exerts against the runway gets converted into a retarding force.

The total force of the tires against the runway is the weight minus the lift, sometimes regarded as the *net weight*. After the nose is lowered on tricycle gear aircraft, the lift is pretty much dissipated and the net weight is the total weight at landing. Friction coefficient then represents how much of the weight becomes a retarding force. The Greek symbol $\mu$ (pronounced mew) is usually used to denote the coefficient of friction. Frictional force is $\mu$ times the net weight.

$$F_{frict} = \mu \times W_{net}$$

The braking capacity of most airplanes produces a coefficient of friction of from 0.4 to 0.7, meaning that as much as 70 percent of net weight can be converted to frictional force; however, this figure is only for dry concrete and also represents a condition where the wheel is not entirely locked. Interestingly, a fully locked wheel is not as effective in braking action as one allowed to roll somewhat. The most effective case is where the brakes are applied to reduce the roll only approximately 10 percent more than what it would be with no braking. Locking the wheels will reduce a braking friction coefficient from 0.7 at 10 percent rolling to about 0.5 in the fully locked condition. Rain, ice, and snow also greatly affect the braking ability. The following list shows the resulting values of $\mu$, which is a maximum of 0.7 for a dry concrete runway:

| Surface | $\mu$ |
|---|---|
| Dry concrete | 0.7 |
| Light rain | 0.5 |
| Heavy rain | 0.3 |
| Snow or ice | 0.1-0.2 |

As a comparison to how effective brakes are in creating friction, the normal rolling friction coefficient of the average tire on a concrete runway without any braking is approximately 0.02.

Other means of deceleration can also be applied. Reverse thrust is the most practical of these and can be achieved by reverse pitch on propeller-driven aircraft, or by deflecting the flow forward in the case of jets. Drag chutes are also employed on high-performance military aircraft, but their use, although effective, is involved and not very practical for commercial aircraft—particularly light aircraft.

Another way of shortening the overall landing distance over a 50-foot obstacle is to reduce the airborne portion (that required for descent from 50 feet). Recall that glide ratio is equal to L/D ratio, so that low L/D means low glide ratio or steep descent. A steep descent results in a short horizontal distance for a given altitude loss; thus, low L/D reduces the airborne distance and is achieved by high drag. Usually we seek low drag to yield high L/D for efficient flight. The landing approach is one exception to that rule. A dirty configuration here is favorable to a shortened approach.

One common way of increasing drag is use of flaps; however, flaps also increase lift, which seemingly would increase the L/D ratio. Flaps actually increase lift *coefficient*, so that for a given weight, the aircraft can be operated at a lower airspeed. The maximum lift coefficient is also increased with flaps, so that the stall speed is reduced. The result is a slower possible touchdown speed. Slower touchdown, in turn, means a shorter range of speed through which to decelerate and the landing roll distance is reduced. Flaps have the effect of reducing airborne distance and ground roll distance. They are a friend indeed to landing performance.

Landing is affected by wind and runway slope in just the same way that takeoff is affected. Altitude (pressure and temperature) also affects landing performance, but not as much as takeoff because engine power is not involved. Weight affects the airborne distance because increased weight means higher L/D and thus longer airborne distance. The ground roll is not affected by weight, though. Higher weight means more mass to be decelerated, but also a proportional amount of increase in braking force, due to frictional force being proportional to weight; thus, the two factors cancel each other out, and the ground roll is unchanged.

# JET AIRCRAFT PERFORMANCE

Jet-propelled aircraft produce thrust directly from the engine. No propeller or other thrust-producing device must be powered as an intermediate stage. The fuel consumption of a jet engine is directly proportional to the amount of thrust that it develops; therefore, jet performance is more conveniently measured in terms of thrust rather than power.

## Speed and climb

For jet aircraft, the drag curve is often referred to as the *thrust required curve*, analogous to the power required curve of propeller-driven aircraft. The jet engine thrust output is then called *thrust available* and compared to the thrust required to determine performance.

Figure 5-23 shows a typical plot of thrust available, $T_A$, and thrust required, $T_R$, for a jet. The thrust available is shown as constant for all velocities, which is approximately true in most cases. The maximum velocity occurs at the intersection of these curves, just as it does at the intersection of power curves. The thrust required corresponding to any given velocity below this point represents the amount of thrust that the engine must deliver for straight and level flight at that velocity.

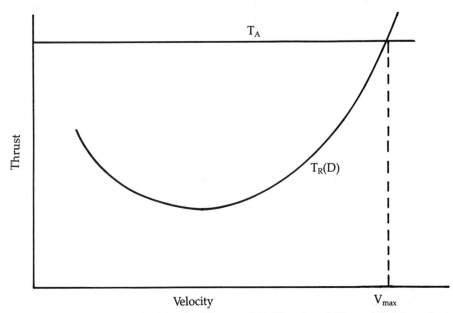

Fig. 5-23. Jet performance is determined from a plot of thrust available and thrust required; the maximum speed occurs where these curves cross.

Rate of climb for a jet is proportional to excess thrust—that is, thrust available minus thrust required—in much the same way as it is proportional to excess power for propeller craft; however, jet climb is also proportional to airspeed so that the speed for maximum rate of climb does *not* correspond to the speed for maximum excess thrust. Rate of climb for a jet is actually equal to excess thrust times velocity divided by weight.

$$RC = \frac{(T_A - T_R) \times V}{W}$$

Therefore, maximum rate of climb occurs at the point where the whole term $(T_A - T_R) \times V$ is a maximum.

## Range and endurance

Because fuel consumption is proportional to thrust, the minimum fuel consumption (lb per hour) would occur at the minimum thrust required; therefore, best endurance occurs at the minimum point on the thrust required (drag) curve.

The speed for best endurance, $V_E$, is shown on Fig. 5-24. For level flight performance, where lift equals weight, this is also best L/D speed.

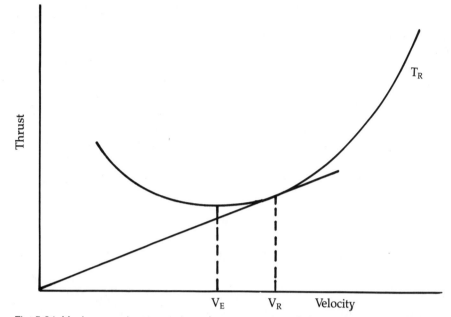

Fig. 5-24. Maximum endurance and maximum range speeds for a jet occur at the same position on the thrust required curve as the same speeds do on a power required curve for a propeller aircraft.

Best range for a jet, however, occurs where the ratio of $T_R$ to V is a minimum. This point is where a straight line drawn from the origin of the graph just touches tangent to the thrust required curve. The speed at this point is called $V_R$, and is also shown on Fig. 5-24. Note that $V_E$ occurs at the minimum point of the curve and $V_R$ at the tangency point, but this is the *thrust* required curve. $V_E$ and $V_R$, maximum endurance and maximum range speeds, for a propeller airplane are at the corresponding points on the *power* required curve.

Because the thrust required curve is actually the drag curve, the minimum point on the curve represents minimum drag, and hence, maximum L/D for a given weight. Because L/D is equal to glide ratio, maximum glide ratio would also be obtained at the speed for minimum drag. It can be shown that this speed also gives the best *angle* of climb; therefore, in the case of a jet, the speed for maximum L/D ratio gives the maximum performance in the following items:

- Endurance
- Power-off glide ratio
- Angle of climb

Best range is obtained at the speed for minimum $T_R/V$ (or D/V) ratio, and best rate of climb at the speed yielding the highest value of $(T_A - T_R) \times V$.

# Takeoff

The takeoff of a jet, or any aircraft certificated under FAA Part 25 (Transport Category), is a bit more complicated than the takeoff of a light airplane. Selected significant speeds during the takeoff are explained under the section on balanced field length. Part 25 also defines other speed requirements that must be observed. Figure 5-25 shows a jet takeoff progression.

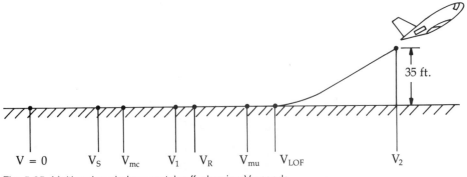

Fig. 5-25. Multiengine airplane on takeoff, showing V speeds.

The airplane is actually capable of flying once it exceeds the stall speed in the takeoff configuration; however, additional speed is required for various aspects of safety. With multiengine aircraft, if one engine fails, there is a yawing moment created by the greater thrust of the remaining operating engines, usually located on the other wing or side of the fuselage. This yaw moment must be counteracted by opposite rudder, if the airplane is to remain headed down the runway. Slower than a certain speed, there is not enough aerodynamic force on the rudder to create sufficient correcting torque. This speed is called *minimum control speed*, $V_{mc}$. The airplane must remain on the ground until this speed is reached. Recall that $V_1$ is the critical engine failure speed, now referred to officially as *decision speed*. This speed must be at least as fast as $V_{mc}$.

Rotation speed is the proper speed to start rotating the airplane for the angle of attack necessary for liftoff. This speed is known as $V_r$, and must be at least 5 percent faster than $V_{mc}$; however, rotation speed need not be any faster than $V_1$. Speed continues to build, reaching a point where the pilot could take off if the maximum possible rotation angle were achieved; this is when the tail actually contacts the ground. This speed is referred to as *minimum unstick speed*, or $V_{mu}$. Because this technique would be annoying to the passengers, to say the least, the actual liftoff is made with some margin of speed above this point. *Liftoff speed*, $V_{lof}$, must be at least 10 percent above $V_{mu}$ for all engines operating, or 5 percent with one engine out. Once airborne, the airplane accelerates to a safe takeoff climb speed, $V_2$, which must be at least 20 percent above stall and 10 percent above $V_{mc}$. This speed must be reached at the obstacle clearance height, which is 35 feet, in the case of Part 25 aircraft.

To summarize, the various takeoff speeds are listed below, in the order in which they normally are reached.

$V_s$    stall speed in takeoff configuration
$V_{mc}$   minimum control speed for one engine out
$V_1$    decision speed for one engine out (= or > $V_{mc}$)
$V_r$    rotation speed (5 percent > $V_{mc}$)
$V_{mu}$   minimum unstick speed at which safe flight is possible
$V_{lof}$   proper liftoff speed (10 percent > $V_{mu}$ all; 5 percent 1 engine out)
$V_2$    takeoff climb speed to be reached at 35 feet altitude (20 percent > $V_s$; 10 percent > $V_{mc}$)

# MANEUVERING

Thus far we have treated all in-flight performance in the unaccelerated condition. In other words, it was steady flight performance; in such case all forces in Fig. 4-1 are in balance. The lift is equal to the weight and the thrust is equal to the drag. Suppose, now, that the pilot pulls back on the stick rather abruptly. The airplane will nose up to a higher angle of attack, and more lift will be created. Momentarily, the lift will be greater than the weight, and the pilot (and passengers) will experience a force tending to push the body down into the seat. This is the reaction to the greater lift force, and is usually expressed in so many g units, or multiples of the acceleration due to gravity.

It should be noted that in light aircraft, where the thrust is much less than the weight, this acceleration will only be momentary. At the higher angle of attack, the drag will also increase and cause the airplane to slow down and reduce the lift to equal the weight. In fighters and other high-performance aircraft, this situation might be different. A more sustained acceleration can also be experienced with light aircraft in a pull-up following a dive.

In any case, when the acceleration, or g-force is imposed upon the airframe, it puts a higher load on the structure than the 1-g force of steady level flight. Obviously, the structure must be designed to withstand more than the steady flight loads, but it cannot be made too strong or the aircraft will have unnecessary weight; therefore, some practical limits must be set. The FAA establishes these limits by specifying certain *load factors* for each category of aircraft. A load factor is the maneuver force in a particular direction divided by the weight of the aircraft. In the vertical direction, the load factor is the lift divided by the weight. The load factor is given the designation, n. The relationship is written mathematically,

$$n = \frac{L}{W}$$

The lift in an accelerated condition then equals n times the weight. An airplane in a pullup will experience lift equal to nW, as shown in Fig. 5-26, rather than W, as was indicated on Fig. 2-1 for steady flight.

While the occupants are experiencing a downward push, the wings of the airplane are being pulled up with the same amount of force, due to this increased lift. A wing of an aircraft pulling 2 g's, for example, will have twice the loads nor-

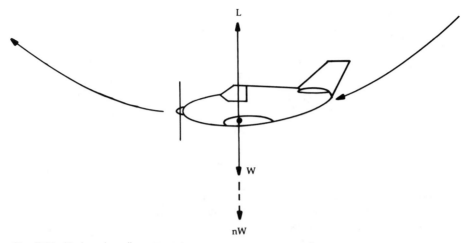

Fig. 5-26. Airplane in pull up maneuver.

mally imposed in steady flight. Limit loads that an aircraft must be designed to withstand are in Table 5-1, vertical load factors. Load factors that result from a pullup are positive; load factors that result from a pushover are negative; load factors with a range of values depend on gross weight.

**Table 5-1. Vertical limit load factors from FAA Part 23 and 25.**

| Category | Positive | Negative |
|---|---|---|
| Standard | 2.5 to 3.8 | 1.0 to 1.52 |
| Utility | 4.4 | 1.76 |
| Acrobatic | 6.0 | 3.00 |
| Transport | 2.5 to 3.8 | 1.0 |

A *V-n diagram* is the aircraft operational envelope that ensures design loads are not exceeded; the diagram is simply a plot of velocity for various load factors; occasionally the plot is called a *V-g diagram* because load factors are often expressed in g units. Figure 5-27 shows a typical V-n diagram for a light airplane. The airplane is certificated in standard category; hence, the limit loads are +3.8 (for this weight) and −1.52. These values make up the upper and lower limits of the diagram.

Stall speed, $V_s$, according to the takeoff section of this chapter, is proportional to the square root of the weight, divided by the wing area and the maximum lift coefficient. This relation was for steady flight, so that the weight was substituted for the lift. In accelerated flight, the lift, or upper term in this equation, is nW; thus, the general relationship is,

$$V_s \propto \sqrt{\frac{nW}{SC_{L_{max}}}}$$

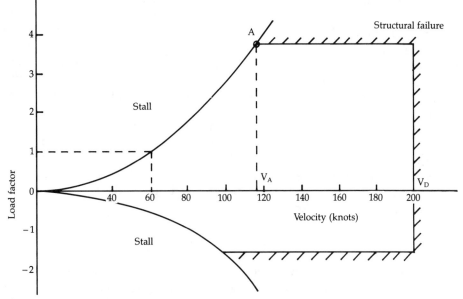

Fig. 5-27. Velocity-load factor (V-n) diagram.

Hence, at 1 g the upper term is W, but under more forceful g conditions, the n is greater than 1, and the stall speed would then be faster. The stall speed would actually be equal to the square root of n times the stall speed at the 1-g condition.

$$V_s = \sqrt{n} \times V_{s1}$$

Obviously, the airplane cannot fly at a speed slower than stall, so that a line representing the stall speed for any n (g-force) becomes a lower limit on our V-n diagram. This could easily be computed by calculating the 1-g stall speed, and then multiplying it by the square root of n for various n values.

For our example, let us assume that the 1-g stall speed is 60 knots. The stall speed for 2 g's would be the square root of 2 times 60, or 85 knots. Computing several other values similarly, we could then plot the slower speed end of the diagram. The part of the curve below zero is done similarly using the inverted stall speed values. The 1-g stall speed is higher here because a cambered airfoil (used in most wings) has a much lower minimum (maximum negative) lift coefficient than in the normal positive direction.

The other end of the diagram is the maximum allowable speed. Again, like the n limits, this must be specified by regulation. It is referred to as the maximum dive speed, or $V_D$. For our airplane, let us assume that this value is 200 knots. This completes the envelope.

At load factors higher than +3.8 or −1.52, or at speeds faster than 200 knots, structural failure could occur. At speeds below the curves on the left side of the diagram, flight is not possible because the airplane would be stalled.

An important corner of the envelope is the intersection of the stall limit curve and the positive limit load factor, point A on the diagram, and the corresponding velocity is called $V_A$. At any speed slower than this velocity, it is impossible to pull more than the maximum g's for which the airplane is designed. In flight operations parlance, this is called *maneuvering speed*, and represents the maximum speed to use during maneuvers, in turbulent air, or any other situation where abrupt control inputs are anticipated. For our example airplane, maneuvering speed is 117 knots.

Note that this stall limit line was calculated for the airplane at full gross weight. At any less weight, this curve would move to the left, meaning that the maneuvering speed would *decrease*; thus, for most operating weights, the maneuvering speed will be slower than that specified for the design gross weight. This fact suggests that a significant margin be applied to maneuvering speed under severe conditions.

## ACCELERATED CLIMB

Climb performance described previously assumed climbing at a steady forward speed. Such performance is typical of light aircraft. High performance airplanes, however, with high thrust-to-weight ratios, can actually accelerate during the climb. In many cases, a rather wide variation in speed will produce a significant increase in climb rate.

Rate of climb was previously defined as the excess power, P, divided by the weight, W, or excess thrust times velocity divided by weight, in the case of a jet. These relations are true only if there is no acceleration, or change in forward speed; that is, climbing at a steady speed. The excess power divided by the weight (or excess thrust times velocity divided by weight) is sometimes referred to as *specific excess power*. "Specific" is used in technical language to denote some quantity per unit of another. In this case, it is the excess power per pound of weight of the aircraft. Specific excess power is given the notation $P_s$.

Specific excess power can be used to climb, or it can be used to accelerate; hence, for higher powered aircraft, there is a trade-off, and the excess power can be used partially for climb and partially for acceleration. Mathematically, it works out to the following equation:

$$P_s = RC + \left( \frac{V}{g} \times a \right)$$

where,

$P_s$ = specific excess power
RC = rate of climb
V   = forward speed
g   = acceleration of gravity
a   = acceleration or change in speed per unit
    of time (usually per second).

For example, a jet climbing out at 175 knots after takeoff, with a maximum rate of climb of 3,000 ft/min at that speed, could maintain a steady velocity of 175 knots and use all of the excess power for climbing at 3,000 ft/min. If, on the other hand the pilot needed to accelerate, he could level off and use all the excess power to gain speed. In this case, the aircraft would accelerate, initially, at 5.5 ft/sec (3.25 kts) each second. The acceleration rate continually changes because it depends on the forward velocity, which, of course, is now increasing. The point is that the specific excess power can be used for either vertical or forward velocity, or some combination of these. Climbing at 2,000 ft/min, for example, would allow for sufficient $P_s$ remaining to simultaneously accelerate at 1.8 ft/sec/sec.

Figure 5-28 shows a jet on the airborne portion of the takeoff run. Notice that for about the first half of the horizontal distance, nearly all the excess power is used for acceleration, with very little climb. The last half of the run is then used to climb to the 35-foot obstacle clearance height, with very little acceleration. This is typical of a jet takeoff, and usually provides the best way to reach $V_2$ at the required height in the shortest horizontal distance.

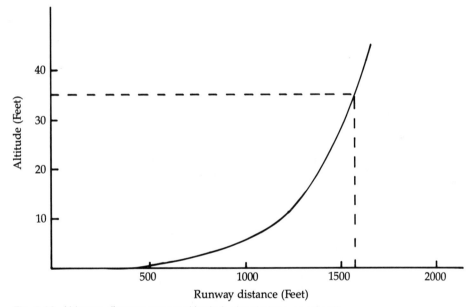

Fig. 5-28. Airborne distance versus altitude for jet airplane on takeoff.

Accelerated climb performance is usually described from an energy standpoint. The potential energy of any body is its weight times the altitude, and the kinetic energy is one-half the mass times the velocity squared. Using these principles, the stick and rudder crowd defines a quantity known as *energy height*, $H_e$, also known as *specific energy* because it is also per pound of aircraft weight. It is the actual altitude plus the velocity squared divided by two times the acceleration of gravity.

$$H_e = h + \frac{V^2}{2g}$$

Figure 5-29 shows various energy heights plotted on a graph of altitude versus forward speed. Consider an airplane at point A. It has an airspeed of 425 knots at an altitude of 12,000 feet, but also an energy height of 20,000 feet. This means that the airplane has sufficient energy to zoom to an altitude of 20,000 feet and zero velocity (point B), or to dive to zero altitude and achieve 672 knots at point C. This is just a way of portraying the possible tradeoff between kinetic and potential energy. Note, also, that an airplane at point D has a greater energy height than the airplane at A, even though it is lower in actual altitude; however, it has much higher velocity that gives it greater overall energy.

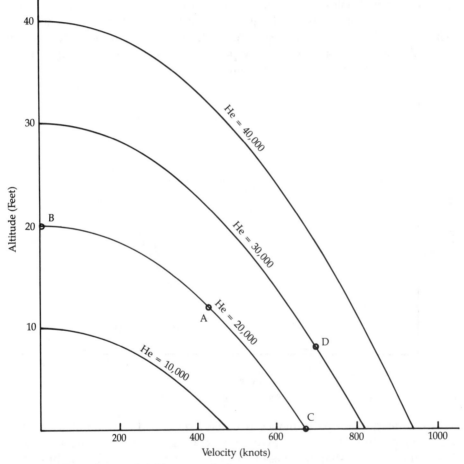

Fig. 5-29. Chart of energy height versus velocity.

This plot of energy height is universal and does not depend on any characteristics of an individual airplane, similar to the variation in temperature or pressure in the standard atmosphere. What gives the aircraft its particular performance capability is its specific excess power, which depends on the particular combination of thrust, drag, and weight for that aircraft. Curves of constant excess power for a typical airplane are plotted on the same energy height graph in Fig. 5-30. The highest curve represents zero excess power, which means that there is just enough thrust to equal the drag, and, hence, maintain level flight. At lower altitudes the curves represent various degrees of excess power over that required for level flight.

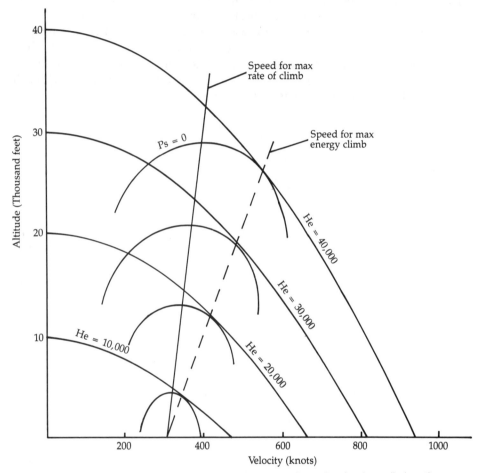

Fig. 5-30. Specific power curves plotted on energy height chart, showing best climb paths.

The minimum time to climb to a particular altitude would be to follow the path through the maximum points on the $P_s$ curves, shown by the solid line. Note that these points are all at a rather slow airspeed. The airplane would then have to accelerate at a constant altitude to reach normal cruising speed. The fastest way to arrive at the desired cruising altitude and cruising speed would be to climb to the corresponding energy height in the least amount of time. This feat is achieved by climbing through the points where the curves of constant $P_s$ are tangent to the lines of constant energy height, shown by the dotted line.

This method of optimum energy climbing was developed in the early years of jet fighter operation. It is particularly useful for military pilots to gain the maximum advantage over enemy aircraft; however, it is also useful in commercial operations to optimize the cruise-climb performance. With the advent of supersonic fighters, it proved to be even more advantageous. As an aircraft approaches Mach 1, the drag rises considerably (see Fig.7-7), and then drops off again in the supersonic range. This causes an irregularity in the $P_s$ curves, and often necessitates the airplane to actually dive through the transonic region in order to maximize rate of climb. The climb schedule for some supersonic fighters can be quite complex, involving both dives and zooms at various stages of the climb. Such maneuvers require detailed instructions in the cockpit, much like approach procedures or aerobatic routines. They can, however, reduce the time to climb to very high altitudes by as much as 50 percent.

# 6

# Stability and Control

A CAMBERED AIRFOIL CANNOT CREATE LIFT and remain in equilibrium, as noted in chapter 2. The very process of creating lift by lowered pressure over the airfoil surfaces also creates a nose-down moment on the airfoil as shown in Fig. 2-23. Some auxiliary device must be provided to the wing to counteract this tendency in order to allow it to fly. The horizontal tail surface is the device normally relied on for this purpose. Conventional tails are placed behind the wing and set at slight negative angles with respect to the wing chordline. This arrangement gives a downward lift force on the tail and, with this tail at a considerable distance (*moment arm*) from the center of gravity, results in a balancing nose-up pitching moment. Figure 6-1 shows a horizontal tail developing a moment to balance out the nose-down pitching moment of the wing when the CG is at the aerodynamic center of the wing. The aerodynamic center and the principle of moments are discussed in chapter 2.

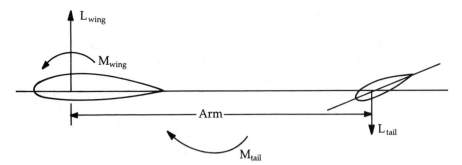

Fig. 6-1. Downward lift on horizontal tail balancing out the downward pitching moment of the wing.

Aeronautical groups usually refer to the balanced condition as *trim*. Trim, or balance, means that no net moment exists that tends to move the airplane out of this condition. We normally think of trim as being established by use of a trimming device such as a tab or movable stabilizer; however, trim in the aerodynamic sense simply means a lack of pitching moment and could be achieved either by tab adjustment or by the pilot's muscle power on the stick. The requirement for balance or "trimability" is absolutely essential to flight.

*Stability*, in the true sense, refers to the airplane's tendency to return to the trimmed condition after a disturbance. Airplanes can have varying degrees of stability, or even be *unstable* and still fly. Many modern fighters are inherently unstable.

The requirement for stability and balance in aircraft was first recognized in the early 1800s. Sir George Cayley is the first aviation pioneer to have recorded any reference to this important aspect of flight. Henson and Stringfellow, English experimenters, used tails behind the wing in their designs of the 1840s. A French engineer named Alphonse Penaud is credited with developing the theory for stability that required a negative tail incidence. He successfully flew a model with such an arrangement in 1871.

By the late 1800s, the need for stability was well known, Chanute, Lilienthal, and Langley all designed their aircraft with great consideration given to this quality. The Wright brothers, paradoxically, recognized the importance of stability, but designed their machines to be unstable. They also recognized the need for control, which is easier to achieve when the airplane is not too stable. Their approach relied on the pilot to provide stability through the controls, thus making maneuverability much easier; however; an inherently stable airplane is a much safer airplane and this quality is required in modern commercial aircraft.

The Wright brothers also placed the horizontal tail ahead of the wing, rather than behind. Such arrangement is referred to as a *canard* configuration, and requires an upward lift on the surface to provide a nose-up moment. Canards have some advantages and some disadvantages that are discussed in greater detail in chapter 10. The primary reason for the Wright canard was for protection against stall accidents. The brothers were concerned when Lilienthal died as a result of a stall during one of his glider test flights.

## THE MEANING OF STABILITY

Before considering the stability characteristics of airplanes, we should first examine the meaning of stability in general. Stability, or the lack of it, is a characteristic that exists in many systems of which our world is composed. The classic example of stability is the small ball in a saucer. Figure 6-2A shows such a ball at equilibrium in the bottom of the saucer. If the ball is moved by some external disturbance to either side (such as the position shown by the dotted outline), it tends to move back toward its original position. This situation represents a *stable system*. The ball seeks its original condition (*equilibrium*), moving back against the direction of the disturbance.

Now consider the case where the ball is placed on the bottom of an inverted saucer, as shown in Fig. 6-2B. Initially the ball is again in equilibrium. It remains at rest as long as no outside forces are brought against it; however, if a disturbance were to push it to one side, as shown by the dotted outline, the ball would not tend to return to its original position, but move farther away. It tends to con-

Fig. 6-2. Stable ball tends to return to equilibrium position in bottom of saucer if displaced to dotted position (A), while unstable ball rolls away if displaced from equilibrium on inverted saucer (B).

tinue moving in the direction that it is pushed. This situation represents an *unstable system*.

Pilots experience instability in air masses in the form of turbulence. A horizontal flow of air encountering an area where warmer air is in the lower layers of the atmosphere will be pushed on upward, if disturbed in that direction. The warmer air below and colder air above are seeking to displace each other because the colder air is heavier. Vertical movement of the air, or *thermal activity*, is thus induced and results in a rough trip for airmen flying through it.

## AIRPLANE AXES

An airplane operates in a three-dimensional environment; therefore, it can rotate about three axes. Stability will be exhibited about all three of these axes and each case is usually considered individually. Figure 6-3 shows the three axes of an airplane and the motion due to moments about these axes.

Movement about the *lateral axis*—that is, nose-up or nose-down—is called *pitch*; rotation about the *longitudinal axis* is called *roll*; and motion due to a moment about the *vertical axis* is called *yaw*. These terms are fairly standard throughout aviation circles, and should be familiar to most pilots. It is important to note that all of these axes pass through the center of gravity (CG). When the airplane is in flight, any force exerted that tends to make the craft rotate will result in rotation about the CG. This is why center of gravity location is so important in aircraft operations.

Stability about each of the three axes is not referred to by the axis of rotation; stability about the lateral axis (pitching motion) is referred to as *longitudinal* behavior because it involves rotation of the airplane as it moves along the longitudinal axis; stability about the roll axis is called *lateral* stability; and yawing characteristics about the vertical axis are referred to as *directional* behavior. Airplanes are therefore considered for their *longitudinal stability*, or stability in pitch; *lateral stability*, or stability in roll; and *directional stability*, or stability in yaw. Lateral and directional stability are not entirely independent of each other, and have to be considered together for the determination of certain motion. Remember that a turn involves both roll and yaw. It is usually not too difficult to make an airplane stable laterally and directionally. By far, the most important consideration in airplane design for stability is longitudinal stability.

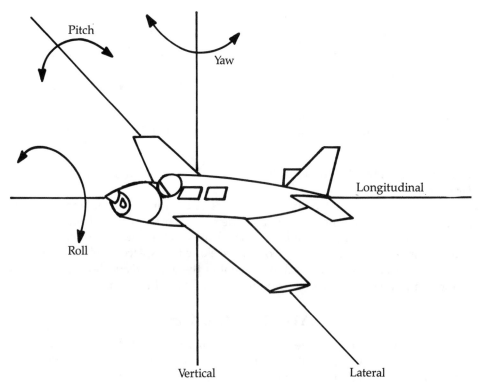

Pitch

Yaw

Longitudinal

Roll

Vertical

Lateral

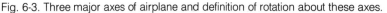

Fig. 6-3. Three major axes of airplane and definition of rotation about these axes.

# LONGITUDINAL STABILITY

An airplane is in equilibrium when it is flying along with all forces in balance, as shown in Fig. 4-1. Even though it might be moving across the ground at several hundred miles per hour, it is doing so at a steady rate. It is not accelerating—that is, changing speed—in any direction. Equilibrium is achieved when no unbalanced force tends to move the airplane (or any object) out of its present state. The airplane in steady flight is then every bit as much in equilibrium as the ball in the bottom of the saucer. The airplane does not necessarily have to be at rest, as in the case of the ball.

Suppose that as the airplane moves along it encounters a pitch disturbance that does move it out of equilibrium. Such a disturbance could be caused by a vertical gust, a CG shift, or a pilot moving the controls. Regardless of the cause, let us assume that the result is a nose-up pitching moment on the airplane. In order to qualify as stable, the airplane must display a tendency to move back toward equilibrium. To do so the airplane must produce a nose-down pitching moment. If, on the other hand, the airplane tended to increase in its nose-up motion, it would display an unstable condition. This initial tendency for the airplane to restore itself to equilibrium after a pitch disturbance is *static longitudinal stability*. Equilibrium for an airplane in flight is usually referred to as the *trim condition*. Trim, remember, does not necessarily refer to the position of the trimming mechanism. It simply means that the airplane has no net moment acting to rotate it.

Figure 6-4 shows a plot of the pitching moment versus angle of attack for a statically stable airplane. The airplane is in trim at an angle of attack ($\alpha$) of $3°$. Notice that the pitching moment is zero at this point, the requirement for a trim condition. Now suppose that the pilot pulls the nose up to $5°$. At this angle a negative, nose-down, pitching moment is created that tends to rotate the airplane back to the trim point. Conversely, a decrease in angle of attack, or nose-down maneuver, as shown by the reduction to $1°$, would create a positive pitching moment; therefore, for a stable airplane the pitching moment must decrease with increasing angle of attack. The curve of pitching moment versus $\alpha$ results in a line with negative slope for positive stability.

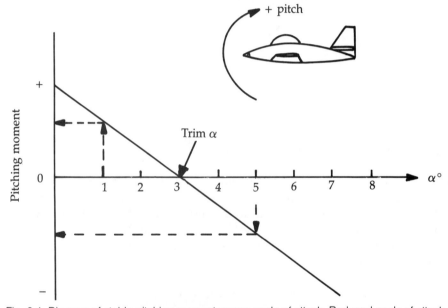

Fig. 6-4. Diagram of stable pitching moment versus angle of attack. Reduced angle of attack produces positive (nose up) pitching moment; increased angle produces negative (nose down) moment.

Different airplanes vary in their *degrees* of stability. Transport aircraft are more stable than most general purpose lightplanes, and most lightplanes are more stable than fighters, for example. The degree of longitudinal static stability is indicated by the degree of slope to the pitching moment curve. Figure 6-5 depicts the static stability of four airplanes. Airplane A exhibits the most negative slope, which makes it the most stable. Airplane B is stable but not as much so as airplane A. The airplane illustrated by curve C is unstable because its pitching moment has a positive slope, or increases with increasing angle of attack.

Curve D represents the stability of an airplane in a unique situation. This curve has neither positive nor negative slope. The pitching moment is constant (the same) at all angles of attack. Such an airplane is said to have *neutral stability*.

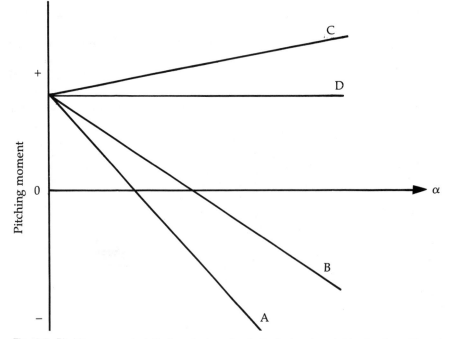

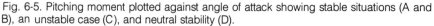

Fig. 6-5. Pitching moment plotted against angle of attack showing stable situations (A and B), an unstable case (C), and neutral stability (D).

If the nose is rotated up to an increased angle of attack there is no moment created either to increase it more or to restore it to trim position. It tends to just remain in the new position. Neutral stability would also be displayed by the example of the ball if it were on a perfectly flat surface. Moving it to one side or the other would create no force tending either to move it toward or away from the original position.

## DYNAMIC STABILITY

Let us again consider the ball in the saucer shown in Fig. 6-2A. If you did move the ball to the displaced (dotted) position and then let go, it would not go directly to the equilibrium point and stop. Rather, its inertia would carry it well past the bottom and up the other side. As gravity overcomes its inertial acceleration, it would roll back the other way toward the original displacement point until it succumbed again to gravity. This motion would go on indefinitely if there were no friction present between the ball and the dish. Eventually, friction would reduce the momentum of the ball sufficiently so that it would finally settle down once again in the bottom. The same action would be displayed by a pendulum pushed to one side and released, or a mass on a spring pulled and then let go. The resulting motion is an *oscillation* (movement from one side of the equilibrium point to the other) until friction damps out the motion.

The airplane will also display an oscillatory motion after a disturbance. If the stick is pulled back until a higher angle of attack is reached and then let go, the airplane will not simply return to the trim angle. If the airplane is statically stable, it will *start* in that direction, but, just like a pendulum, it will nose down past the trim angle and then start back up. New forces and moments begin to act on the airplane as this motion takes place. These influences could cause the oscillations to damp out, as in the case of the free-swinging pendulum, or they could cause the oscillations to increase in amplitude with time. How the airplane behaves over a period of time after a disturbance is the *dynamic stability* characteristic.

Figure 6-6 shows the possible dynamic behavior of an airplane after a pitch disturbance. The charts show the pitch angle with the horizon versus time after being increased from the trim point and then released. Figure 6-6A shows a decreasing oscillation with time and represents a dynamically stable case. Figure 6-6B illustrates a dynamically unstable situation where the oscillations increase with time. A dynamically neutral situation can also result, in which the airplane continues to pitch up and down at constant amplitude, as shown in Fig. 6-6C.

Static and dynamic stability are usually desired in an airplane. Static stability does not guarantee that an airplane will also be dynamically stable; however, the airplane must first be statically stable before it will oscillate at all, and thus exhibit any kind of dynamic stability. Possible cases of stability can be summarized:

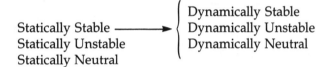

There are also varying degrees of dynamic stability. A dynamically stable airplane will go through a series of oscillations after a disturbance, but these oscillations will gradually reduce in amplitude, as shown in Fig. 6-6A. This means that the maximum pitch angle deviation from the trim angle will continuously decrease until the airplane settles out on the trim angle. The degree of dynamic stability is represented by how rapidly these oscillations damp out. Because it is very difficult to tell exactly when all oscillatory motion stops, a more measurable quality of damping is the time to damp to half amplitude. This is the time required for the oscillations to reduce to half of the original deviation from trim. For example, suppose that the nose is raised 10° from its level flight trim angle and then released. The time to damp to half amplitude would be the time from the release to when the maximum pitch deviation is 5°. The length of time for a typical light airplane is on the order of 20 to 30 seconds. An airplane that damped to half amplitude in 15 seconds would have a greater degree of dynamic stability and one that required 40 seconds, for example, would be less stable dynamically.

The time to damp to half amplitude is formally termed the *long period* (or *low frequency*) mode. It is the only oscillatory motion that a pilot would notice in a typical lightplane. There is also a *short period* (or *high frequency*) oscillation that occurs immediately after the long period motion begins. It damps out quite rapidly, with time to damp to half amplitude on the order of a quarter of a second for

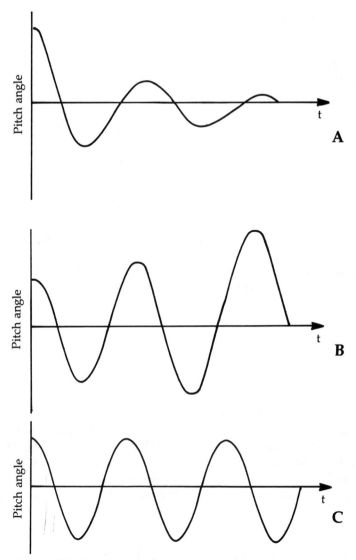

Fig. 6-6. Pitch angle versus time for a dynamically stable airplane (A), a dynamically unstable airplane (B), and a neutrally stable airplane (C).

light airplanes. Such motion is so rapid that it is hardly ever noticed as an oscillation. It may seem to the pilot to be just a slight bump, similar to hitting a small thermal. Short period motion can be significant in higher-performance aircraft and can lead to severe oscillation and high g loads if fed by *pilot-induced oscillation*. Pilot-induced oscillation results when a pilot attempts to counteract the motion, but his reaction time is shorter than the period of oscillation. By the time his control input is felt, it is actually in the direction of the natural motion (which has by then reversed) rather than against it.

# CENTER OF GRAVITY EFFECTS

Figure 6-1 showed that a tail was required to counteract the nose-down pitching moment of the wing. The illustrations neglected any effects of the airplane's weight and center of gravity. The *center of gravity*, CG, is the center of mass of the overall airplane, sometimes described as the point where all of the weight can be considered to be concentrated. The CG of any body can readily be determined by sliding it along a sharp-edged balance point, or fulcrum, until the gravity forces on either side are even and the body remains in balance (Fig. 6-7). The CG of an airplane could actually be determined this way if the structure right at that point could support all of the weight.

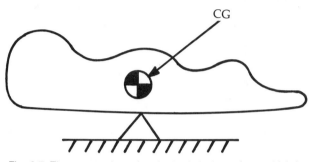

Fig. 6-7. The center of gravity of a body is the point at which it would balance on a fulcrum.

Because that is not usually the case (and also because such an approach would be a bit awkward, to say the least), CG is usually determined by weighing the airplane. The difference in weight between the main gear and the nose or tailwheel and the distance between them can be used to calculate the CG location. Normally this is done for an empty airplane, and loaded CG is calculated by adding weight and moments resulting from additional items such as fuel, passengers, and baggage. Details of such calculations can be found in almost any pilot instruction manual (such as those listed in the bibliography).

A wing can be balanced, trimmed, at a specific angle of attack by proper location of the CG, even without a tail. With the CG located aft of the aerodynamic center, the lift force will give a nose-up pitching moment, as shown in Fig. 6-8. If the distance, l, from the CG to the lift vector is just right to make this moment equal the natural nose-down pitching moment, M, the wing will be in balance; however, the slightest increase in angle of attack will cause the lift to increase by an amount $\Delta L$, and the moments will no longer be equal. The result is a greater nose-up pitching moment. A greater nose-up tendency from an initial nose-up movement (increase in angle of attack) is an unstable situation; therefore, while the wing alone can be balanced with the CG, it cannot also be made stable.

When a tail is added, as shown in Fig. 6-9, balance is obtained partially by the nose-up moment of the wing lift ($L_w$) forward of the CG and partially by the nose-up moment of the downward lift on the tail ($L_t$) at an arm of $l_t$. In this case,

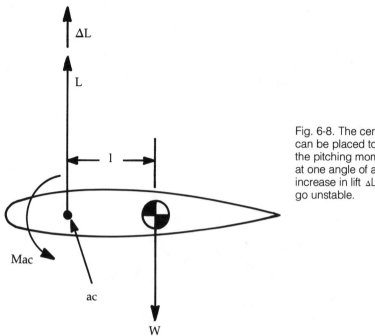

Fig. 6-8. The center of gravity can be placed to balance out the pitching moment of the wing at one angle of attack, but any increase in lift ΔL would make it go unstable.

for trim flight, the natural nose-down pitching moment (moment about the aerodynamic center) is equal to the sum of the wing lift moment and the tail lift moment.

$$M_{ac} = L_w \times l_w + L_t \times l_t$$

Now consider an increase in angle of attack. Remember that the airplane will rotate about the CG. An increase in angle of attack will increase the lift and therefore the nose-up moment, just as it did for the wing alone; however, the tail will now rotate downward, placing it at a *lower* angle of attack, *reducing* $L_t$ and the resulting nose-up moment of the tail. If the moment from the tail is greater than the moment from the increased wing lift, the result will be a nose-down tendency and the airplane will be stable. The horizontal tail is relied on primarily to provide longitudinal stability. The farther aft from the CG that the tail is located, the larger the restoring moment will be. Because this moment is also proportional to tail lift, the moment is also proportional to the horizontal tail surface area. Greater stability results, then, from either an increase in tail area or increase in tail moment arm.

The wing lift, on the other hand, creates a destabilizing moment about the CG. The farther aft the CG is located from the wing's aerodynamic center, the less stable the airplane becomes. In fact, there is a point where the CG could be located so that the moment due to increased wing lift just equals the opposite moment from the decreased tail lift. At this point the airplane would display neutral stability and its pitching moment would be the same at all angles of attack. This is the situation shown in curve D in Fig. 6-5. This point where the CG location would cause neutral stability is the *neutral point*.

It should be obvious now why flight instructors, FAA inspectors, and other advocates of aviation safety keep reminding pilots of the importance of CG location. There has to be some flexibility in CG location because varied loadings must be accommodated to some degree; however, you can see that the farther aft the CG is moved, the less stable the airplane becomes. If the CG were allowed to move all the way to the neutral point, the airplane is no longer stable. If it were to move aft of that point, the airplane becomes unstable. Location of the CG a substantial distance forward of the neutral point is imperative for a necessary degree of stability. It is this requirement that imposes the limitation on allowable aft CG location.

# CONTROLS

Up to this point we have been discussing the balance and stability of the airplane. The implication has been that a desirable situation is one in which the airplane is quite stable. To some degree, that is true. Stability provides the quality of keeping the airplane in its desired flight path; however, sometimes we want to deviate from that path, maneuvering the airplane to climb, descend, or turn. If the airplane is very stable, maneuvering away from its established course might be difficult. It was this quality that the Wright brothers recognized in very stable gliders of their day. They placed so much importance on maneuverability that they designed their airplane to be inherently unstable. We now know that you do not have to go to that extreme, but increased maneuverability does result when the inherent stability of an airplane is decreased in the design. A good compromise between stability and maneuverability is the designers ultimate goal.

## Pitch control

Airplane control is provided by changing the lift on a control surface located at some moment arm from the CG. Figure 6-9 shows the horizontal tail set at some

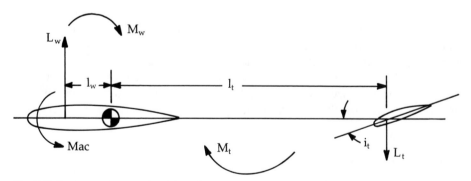

Fig. 6-9. Forces and moments of the wing and tail in balance about the center of gravity and also stable.

angle of incidence ($i_t$) with respect to the wing chord. Recall that this had to be a certain angle to trim the airplane, or balance out the other moments so that the total moment is zero. If, on the other hand, we wanted to create a moment to rotate the nose up or down, we could do so by changing the angle of the tail. Increasing the negative angle of the tail (nose of the stabilizer downward) would increase the negative lift on the tail and rotate the airplane about its CG so that the nose would move up. Conversely, reducing the negative tail angle, or moving the nose of the stabilizer to a positive angle, would reduce the downward lift on the tail and allow the nose of the airplane to rotate downward. Some airplanes are actually controlled this way and the tail surface is then known as an all-movable tail or *stabilator*.

A more conventional means of pitch control is the stabilizer-elevator combination. The elevator acts like a flap, effectively increasing the camber of the tail surface. Increased camber also increases lift, which is the principle on which flaps are based, as discussed in chapter 2. Figure 6-10 shows a stabilizer set at a zero angle of incidence, but the elevator deflected upward at an angle e. This position causes a negative camber to be produced, with a resulting downward lift on the tail. At a moment arm of $l_t$, the overall effect is a moment about the CG tending to rotate the nose upward. Opposite movement of the elevator would yield a positive camber, a positive lift force, and a resulting nose-down moment. Such control places the wing at various angles with respect to the free airstream and is thus the means of controlling angle of attack.

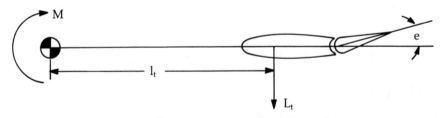

Fig 6-10. Elevator deflection angle (e) causing downward lift on tail and nose-up moment about the center of gravity.

Movement of the elevator or stabilator can also be used to trim the airplane for various flight conditions. Remember that trim means setting the tail at an angle to provide a net zero moment about the CG. The moment required is different for every angle of attack, and, hence, every airspeed. The airplane could be kept in balance at any speed by proper tail angle obtained from corresponding position of the primary control stick; however, this process requires constant force to be exerted on the stick and for long flights could result in a very tired pilot.

## Trim control

In consideration of the pilot, designers have incorporated trimming devices that hold the tail in proper position for trim at the appropriate airspeed without any force on the primary control. One way of doing this is to have an adjustable stabilizer so that the angle of incidence can be set at different angles. This method has the same effect as an all-movable tail, but it incorporates a screw mechanism or other such device that holds the tail in a given position. The elevator will then be at a zero angle in trim. It is used only for maneuvering from the trim speed for short periods of time.

Another device quite common for trimming the airplane with resulting zero stick force is the *trim tab*. With elevator control only, a certain elevator angle is required for trim at a certain speed. If the elevator control were released at any given speed, the elevator would float free to some angle. The airplane would then change speed to match a combination of the free-float angle and the required trim angle. If the elevator were forced to float at the angle required for trim at a particular speed, the stick force felt by the pilot would be zero in that condition. The trim angle is forced into the elevator by use of the trim tab, arranged as shown in Fig. 6-11.

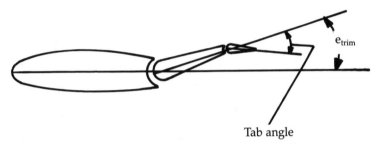

Fig. 6-11. Trim tab deflection forces elevator into required trim angle position.

The trim tab is a movable surface on the elevator, similar to the elevator's relationship to the horizontal stabilizer. In order to place the elevator in the desired trim position, the tab is rotated in the opposite direction. In the example shown, the tab is rotated downward, developing an upward lift on the tab. Because the tab is located at a distance (moment arm) back from the elevator hinge line, this force rotates the elevator up to the trim angle shown. The overall result is a downward lift force on the tail, and a nose-up moment is added to the airplane. With the proper tab adjustment, the elevator moves to the required trim angle for the desired speed, the force on the control stick goes to zero, and the pilot can relax and eat his lunch if flying in stable air. Other methods are employed for trimming, such as a constant force applied through a spring or bungee directly to the elevator.

## Yaw and roll control

The airplane is controlled in yaw (about a vertical axis) by use of the *rudder*. The rudder works just like an elevator, except in a vertical plane. Deflection of the rudder to either side increases the camber of the vertical tail surface and creates a force in the opposite direction. This force, located at a moment arm aft of the CG, causes a moment to act in a horizontal plane about the CG and yaws the airplane.

Roll control also involves a camber change of the wing surface by use of ailerons. Figure 6-12 shows ailerons deflected downward, increasing the camber of the wing over the span of the aileron. This action gives an increased lift, $\Delta L$, in that region of the wing. Because the aileron is located at an arm outboard from the centerline, a rolling moment is induced. The left aileron also moves upward simultaneously, decreasing (or giving negative camber to) the outboard region of the left wing. The result is a downward lift (or decrease in normal upward lift) and a rolling moment to the left is also created. The total rolling moment results from the effects of the deflection of both ailerons.

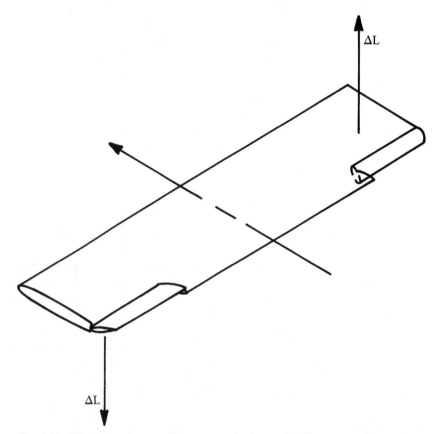

Fig. 6-12. Ailerons produce a rolling moment by increasing lift at some distance from centerline on one wing while decreasing lift an equal distance on the opposite wing.

An interesting aspect of roll control is the fact that in pure roll, there is no opposing moment tending to restore the airplane to a wings-level altitude. The rolling tendency continues as long as the stick is held into the roll and then remains in that attitude when control force is released. This is in contrast to elevator and rudder control, which requires a constant force to be exerted in order to hold the airplane in the desired pitch or yaw position (if not trimmed for that position). Furthermore, in the case of elevator or rudder control, the farther the control is moved, the greater the angular movement (pitch or yaw) will be. In the case of aileron control, the angle of roll continues to increase, so that a large roll angle could eventually result from a slight control deflection. What happens with greater aileron deflection is a more rapid rate of roll. Increased control movement, then, results in greater rate of rotation with ailerons and greater pitch or yaw angle with elevator and rudder.

## DIRECTIONAL STABILITY

The *vertical stabilizer (fin)* provides directional stability to the airplane in the same way that the horizontal stabilizer acts in the longitudinal case. Figure 6-13 illustrates a disturbance that has caused the airplane to yaw to the left. The airplane

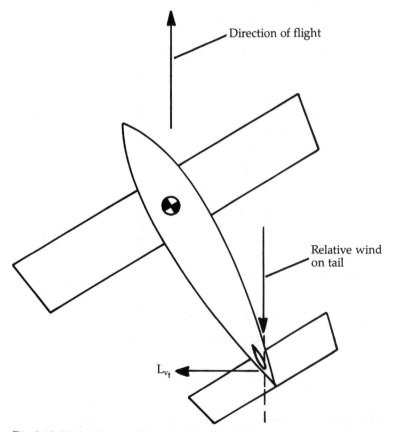

Fig. 6-13. Vertical fin providing directional stability by producing side force in opposite direction of yaw.

also rotates about its CG in this plane; therefore, the tail moves to the right. As it does, the vertical fin is placed at an angle of attack with respect to the relative wind. The result is a lifting force acting to the right on the vertical tail, shown as $L_{vt}$. This side force creates a restoring moment about the CG, swinging the tail to the left and the nose to the right so that it is again lined up with the path of flight. Such stabilizing tendency is also referred to as *weathercock stability*.

The degree of directional stability is proportional to the size of the vertical stabilizer and the distance from the CG. A larger area results in greater side force and a greater distance from the CG yields a longer moment arm. An increase in either of these quantities will produce a larger *restoring moment* (moment is a force times a moment arm). Because the vertical tail is the only significant contributor to directional stability, it must always be located aft of the CG for a stable configuration. Horizontal stabilizers, on the other hand, can be either forward or aft. In the case of the forward mounted horizontal stabilizer (canard), the wing must then be located aft of the CG. In this way the wing provides a stabilizing moment to overcome the destabilizing effect of the canard surface. Because the wing produces no appreciable side force, the task of providing directional stability is handed almost completely to the vertical tail surface.

When yaw is desired, it is obtained by creating a side force on the vertical tail with rudder deflection, as explained in the previous section. The amount of yaw control is proportional to the distance of the tail from the CG (moment arm), the size of the rudder, and the degree of rudder deflection. Sufficient rudder power must be available for turning performance.

Another, often critical, factor in rudder design for multiengine aircraft is the consideration of trim after an engine failure. Because the engines are normally located on the wings, or at some location outboard of the centerline, failure of one engine while the other engine is operating normally results in a yawing moment about the CG. This moment must be counteracted with rudder deflection to create an equal yaw moment in the opposite direction. Like the lift on a wing, side force on the vertical tail decreases with velocity; therefore, even with full rudder deflection, there is a speed below which trimming out the yaw resulting from an engine failure is no longer possible. This speed is called *minimum control speed*, $V_{mc}$. For aircraft certificated under FAR Part 23, this speed must not exceed 20 percent more than the stall speed.

## LATERAL STABILITY

We have seen that if the airplane is rotated in pitch or yaw, stabilizers can provide restoring moments to reinstate the trim position. Pure rolling motion is a different situation. There is really no aerodynamic force created in rolling that tends to restore the wings to level flight. There is also no tendency for the roll to continue once begun, unless it is forced to do so by aileron-induced rolling; however, if aileron deflection is discontinued the wing tends to remain in a bank angle. The airplane can thus be considered neutrally stable in roll. As will be shown, pure roll is not really possible without other motion and resulting effects. The point here is that rolling, in itself, does not produce a restoring moment.

A consequence of rolling the airplane into an angle of bank is to induce a sideslipping in the direction of the bank. When the airplane is banked, the lift vector, always acting perpendicular to the plane of the wings, gets tilted in the direction of the bank. Figure 6-14 shows an airplane banked to the right. The lift vector is also tilted to the right, and therefore has a component in this direction. This component of the lift tends to move the airplane to the right into a sideslip. Means can be designed into the airplane to counteract this sideslipping (lateral) motion. Because this is the primary concern in stabilizing motion about the roll axis, such stability is usually referred to as *lateral stability*.

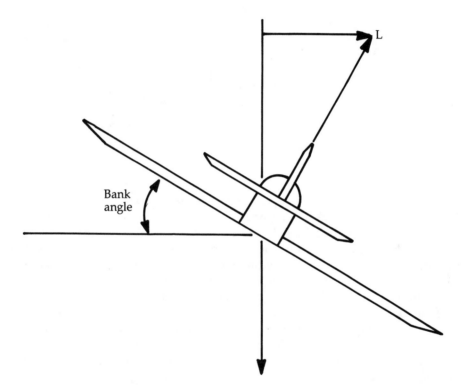

Fig. 6-14. Bank actually causes turning by placing component of lift (always perpendicular to wing plane) in direction of bank.

One of the most effective means of stabilizing the airplane against sideslipping from a roll is to provide *dihedral*, which is tilting the wing so that the plane of the wing is at an angle with the horizontal in the lateral direction. Figure 6-15 shows an airplane with dihedral. If the airplane were sideslipping to the right, as shown, a component of the relative wind would be acting inboard against the right wing. A component of this velocity would be acting against the bottom of the wing, tending to roll it to the left; thus, a roll to the right tends to slip the airplane to the right, but with dihedral, an opposite moment is created to level the wings and arrest the slip.

Some lateral stability is also provided by the vertical tail. As shown in Fig. 6-15, a sideslip also results in a force against the vertical stabilizer. Because the center of the vertical tail is usually above the CG, a moment is created that tends to roll the airplane away from the sideslipping direction.

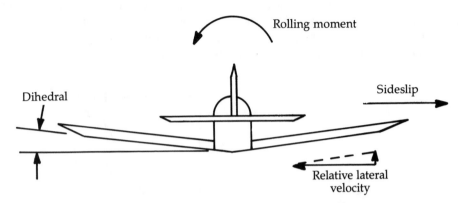

Fig. 6-15. Dihedral provides lateral stability by upward component of relative lateral velocity resulting from side slip.

Another consideration in lateral stability is the fuselage effect. If the airplane is sideslipping, a sideward component of velocity hits the fuselage and flows over the top and bottom. Just ahead of the wing this flow induces an upwash with a high-wing configuration or a downwash with a low-wing on the wing moving into the slip. As illustrated in Fig. 6-16, the upwash effect on the high-wing tends to roll the airplane out of the slip and is thus stabilizing; however, the low-wing encounters a downwash that rolls it farther into the slip, a destabilizing configuration. This explains why more dihedral is necessary in low-wing airplanes. High-wing airplanes often have little or no actual dihedral, yet have lateral stability. The amount of the effect depends on the fuselage cross-section shape and the amount of fuselage ahead of the wing as well as the vertical location of the wing.

Still another contributor to lateral stability is the swept wing. The primary reason for sweeping a wing is to reduce the velocity flowing perpendicular to the wing, and thus reduce the wave drag when operating at speeds near or above the speed of sound. A secondary effect of wing sweep is improved lateral stability. Such effect is so pronounced that it is sometimes referred to as *dihedral effect*. This effect results from the velocity more directly hitting the wing in the direction of the sideslip and less directly on the opposite wing. Figure 6-17 shows the normal direction of the velocity on each wing with forward flight and the direction when sideslip is present.

A more direct velocity on the right wing produces more lift on that wing and the less direct velocity on the left wing produces less lift. The result is a roll to the left, again away from the direction of slip. Some very highly swept aircraft actually have a negative dihedral angle to reduce the dihedral effect of the sweep.

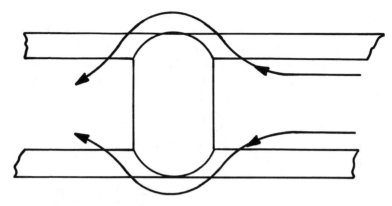

Fig. 6-16. Fuselage gives dihedral effect by producing an upward component of velocity on the wing into the side slip on a high-wing configuration, while the opposite is true on a low-wing.

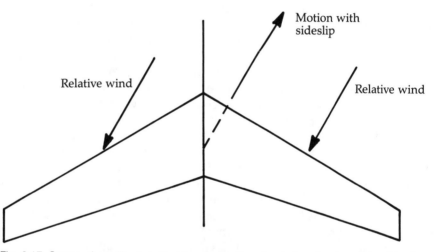

Fig. 6-17. Sweep gives dihedral effect by placing the wing in the direction of the side slip more directly into the wind, producing more lift and rolling away from the slip.

This is particularly true if the airplane also has a high wing, which further adds to the lateral stability. Too much stability reduces the controllability necessary for maneuvering.

# DIRECTIONAL-LATERAL COUPLING

We have been discussing roll and yaw up to this point in the pure sense, as if they occurred independently. In reality, it is impossible to yaw an airplane without inducing some rolling tendency, and, likewise, to roll without some resulting yaw. The interaction between rotating motion in one plane and rotating motion in another plane is referred to as *coupling*.

Perhaps the best known effect involving coupling—especially amongst the helmet-and-goggles group—is *adverse yaw*. If a turn is initiated by aileron displacement alone, the down aileron on the wing on the outside of the turn creates more drag than the up aileron on the inside wing (Fig. 6-12). This action occurs because the wing is a cambered surface and the aileron acts just like a flap over that portion of the wing. The down aileron increases camber and thus increases drag. The up aileron reduces the positive camber and actually reduces drag a bit in the first few degrees of deflection. The results of the uneven drag on the ailerons is a yawing moment, and this moment is away from the direction of the intended turn.

Because the yaw thus produced is in the opposite direction, it is referred to as an adverse situation. If not corrected by rudder, the vertical fin will eventually weathercock, or swing around from a correcting side force produced and end up yawing in the direction of the turn. Such a maneuver is best described as sloppy. Excessive drag is created, sideslipping occurs, and the turning rate is very slow. Again, our ingenious Wright brothers were among the first to recognize this problem, and incorporated rudder control coordinated with roll control (for which they used wing warping) to achieve efficient, smooth turns.

Even without any aileron deflection, rolling motion tends to produce adverse yaw. Such motion might be induced by a vertical gust, for example. During the roll, the upward-moving wing has a downward component of relative wind hitting it. The lift, always perpendicular to the resultant velocity, thus gets tilted slightly aft, shown in Fig. 6-18. Conversely, the downward-moving wing has some vertical velocity component. The resultant relative velocity gets canted upward, and the lift on this wing is tilted forward. The aft component of lift on the rising wing and the forward component on the descending wing causes a yawing moment. Again the yaw is opposite to the intended direction of turn, and contributes to the adverse yaw effect; however, note that this contribution exists only as long as the airplane is rolling. If the wings are held in a constant bank angle, there is no vertical velocity on either wing and the lift on each is in a vertical direction, tending to produce no yaw.

If a turn were begun by rudder deflection, rather than aileron, then the initial motion would be yaw; however, as shown in Fig. 6-19, motion about the yaw axis would result in a greater relative velocity acting on the wing to the outside of the turn and a reduced velocity on the inside wing. Because lift is dependent upon velocity, more lift would be generated on the outside wing and less lift on the inside wing, causing a roll into the turn. Although this effect is into the direction of the intended turn, there is a delay in the rolling and a skid to the outside of the turn would occur. Again, the need for coordinated control is clearly indicated.

If the vertical tail is quite high, so that its center of pressure is well above the CG, the side force produced by the rudder could produce an appreciable rolling moment acting against the direction of turn. This effect can be visualized looking at the side force on the tail in Fig. 6-15. (One of my students suggested that this effect should be termed "adverse roll.") The induced rolling moment is usually

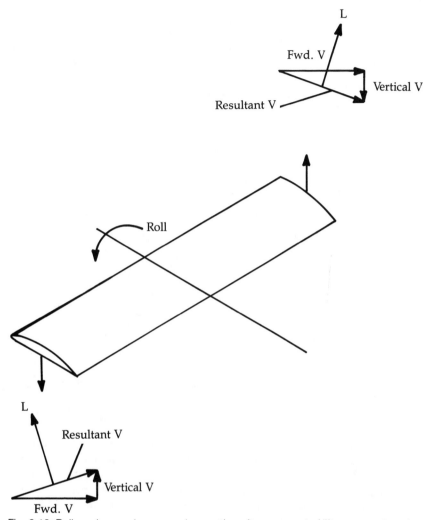

Fig. 6-18. Roll produces adverse yaw by creating aft component of lift on upward-moving wing and forward component on downward-moving wing.

not as great as that resulting from the lift differential on the wings, and the net rolling moment for rudder deflection is into the direction of turn.

## LATERAL DYNAMIC MOTION

Just as in the case of longitudinal motion, an initial tendency to move toward the lateral trim condition does not guarantee lateral stability. When the motion begins, additional forces and moments are created that could either add to or detract from the stabilizing process.

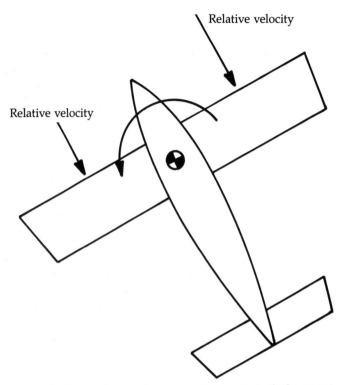

Relative velocity

Relative velocity

Fig. 6-19. Yaw produces rolling by creating greater velocity on outside wing, thus producing more lift.

Because of the coupling effects for roll and yaw, motion about both of these axes would result from an initial disturbance about either one; however, the effects will be different depending upon the degree of rotation in one plane compared to that in the other. If a pure rolling moment could be produced (without any yaw), an opposing reaction takes place with most airplanes, tending to damp out such motion very rapidly.

## Spiral divergence

A very slight yaw motion, on the other hand, usually results in little (or no) resistance to continued yawing. If uncorrected by pilot action, a yaw will usually continue to increase with most airplanes. This increase is very gradual, and in normal flight would be easily recognized and corrected; however, if the pilot takes his eyes away from the instruments or the horizon and concentrates on something else, charts for example, for a length of time, he might be quite surprised when he looks up to see the airplane in a significant turn. This is a result of *yaw instability* and is exhibited by a gradually increasing turn of decreasing radius or spiraling motion.

Many light aircraft are actually unstable in this mode, in which case the motion is referred to as *spiral divergence*. Such instability is allowable under FAA certification as long as the degree of instability is slight. The instability normally creates no problems in flight; however, it can be very dangerous to a pilot without adequate training to fly on instruments. An unchecked spiral divergence also results in loss of altitude and increase in airspeed. If a proper recovery is not initiated, the motion ends up as a spiral dive, sometimes morbidly referred to as a *graveyard spiral*. It can catch the unsuspecting pilot in a dangerous situation primarily because it starts out as a very slight turn that is hardly noticed.

## Dutch roll

A third type of lateral motion that can occur is an oscillating movement from side to side. Such motion can result from a rather large control input that is then suddenly released. A similar effect can be experienced from a sharp gust hitting one wing or the vertical tail. The resulting motion is a series of rolls and yaws that are out of phase with each other, amounting to a series of uncoordinated turns alternately in opposite directions. The motion was once likened to a Dutch boy on skates doing a series of opposite turns. The direction of the turn is always out of phase with the rolling motion of his body (which is how he initiates the turn). This motion in an airplane thus became known as *Dutch roll*.

Most aircraft are stable in this mode. That is, the motion dissipates, or damps out, over a period of time; however, in some airplanes, the damping is slight, and a disturbance will cause the motion to last for a significant period of time. Because it is an uncoordinated motion, it can be very disturbing to passengers. Airplanes with small vertical tails often exhibit significant Dutch roll motion. The Ercoupe was a good example of this. The V-tail Bonanza also has more such motion than its straight-tail sibling.

It is practically impossible to design an airplane to have ideal lateral characteristics in all modes. What improves the airplane in one respect makes it worse in another. In general, increased dihedral (or dihedral effect) will reduce spiral instability, but will increase Dutch roll tendency. Conversely, large vertical tail size tends to reduce Dutch roll motion, but also increases spiral instability. As with all airplane characteristics, design for lateral stability has to be a compromise to give the best overall performance.

## TURNING PERFORMANCE

Most everyone knows that a boat is turned in the water by use of a rudder; hence, it is natural to believe that the airplane is also turned by its rudder; however, an airplane, operating in a three-dimensional medium, is a bit different. Banking of the airplane actually causes it to turn. Because this action is caused by the ailerons, they are the primary turning controls. The rudder is there to provide for a *coordinated turn*. To see how this maneuver works, consider the diagram in Fig. 6-20.

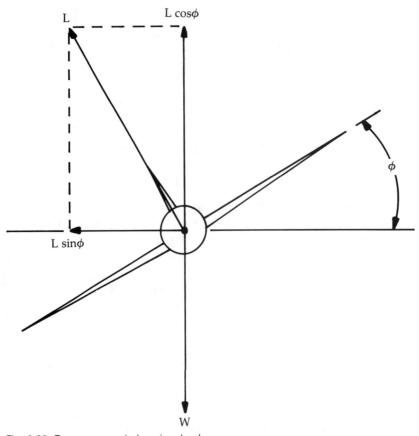

Fig. 6-20. Forces on an airplane in a bank.

## Anatomy of the turn

The airplane is banked through an angle $\phi$ (pronounced fee or fye). Lift, remember, is always perpendicular to the plane of the wing. Because the wing is tilted, the lift vector is also tilted; however, weight always acts downward (toward the center of the earth). Enough lift must be generated for the vertical component of the lift to equal the weight, if the airplane is to remain in level flight. Recall from trigonometry that this component of lift is equal to the lift times the cosine of the bank angle, where the cosine is defined as the side adjacent of the triangle divided by the hypotenuse.

The tilted lift vector also has a horizontal component, lift times the sine of the bank angle, which is tending to pull the airplane into the direction of bank. This pull should result in the airplane just moving sideways, and would, except that the vertical fin provides weathercock stability and thus aligns the longitudinal axis with the resultant motion. Figure 6-21 shows the velocity resulting from the combination of forward and sideward velocities. This aspect of flight is explained in more detail in the section on Directional Stability.

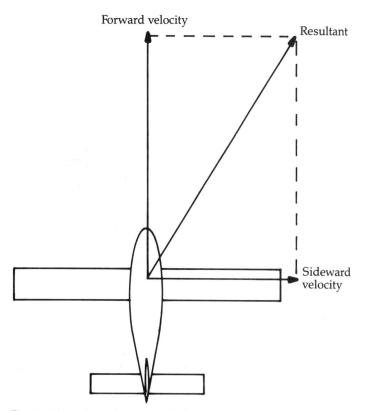

Forward velocity

Resultant

Sideward velocity

Fig. 6-21. Velocity vectors on an airplane in a bank.

Aileron deflection alone, of course, produces adverse yaw, as previously explained; therefore, rudder control is also required in order to achieve a coordinated turn. Also, once the turn has begun, the outside wing is moving at a greater velocity and produces slightly more drag than the inside wing. This effect adds to the adverse yaw tendency, and is another reason that some rudder deflection is needed in the direction of the turn.

## Radius and rate of turn

From Newton's laws of motion, when a body is moving in a circular path, it is subjected to a centrifugal force equal to its mass times the square of the forward velocity at any point, divided by the radius of the circle (Fig. 6-22):

$$F_{cent} = \frac{m \times V^2}{R}$$

Mass can be expressed as the weight divided by the acceleration of gravity, $W/g$. The centrifugal force must be equal and opposite to the inward component of the lift, by Newton's third law, and this lift component is equal to the weight times

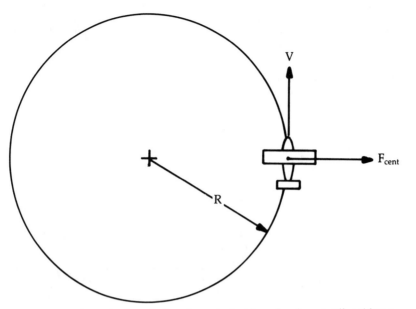

Fig. 6-22. Circular path of an airplane in a constant turn showing centrifugal force.

the sine of the bank angle (Fig. 6-20). Rearranging the equation algebraically to solve for the radius of turn, the weight cancels out, yielding another equation:

$$R = \frac{V^2}{32.2 \text{TAN} \phi}$$

This is the mathematical expression that simply says that the radius of the turn is equal to the velocity squared, divided by the acceleration of gravity, which is a constant term of approximately 32.2 ft/sec$^2$ at lower altitudes, and the tangent of the bank angle. Because the tangent is proportional to the angle, this means that higher bank angle results in smaller radius of turn (because it is in the denominator). Higher velocity, on the other hand, means larger radius (actually very much larger because it is squared); hence, the smallest radius turn is made at the lowest velocity and the steepest angle of bank.

The *rate of turn* can also be shown, by some higher mathematical operations, to be equal to the velocity of the airplane divided by the radius. Because the radius is dependent on both velocity and bank angle, if we use the above expression for turn radius, we can eliminate the radius (usually an unknown) and obtain the turn rate in terms of bank angle. For the rate in degrees per second (the usual measure), with velocity in knots, and using 32.2 ft/sec$^2$ for g, the formula is:

$$RT = \frac{1091 \text{ TAN} \phi}{V}$$

Here, we see that the rate of turn (RT) is proportional to the angle of bank and is inversely proportional to the velocity. This means that rate of turn is greatest at the lowest possible velocity and the highest angle of bank. These are the same conditions that yield the smallest radius.

## Load factor and stall speed

From Fig. 6-20 we can see that only the vertical component of the lift is acting to overcome the weight. Part of the lift is involved in pulling the airplane into the turn. The total lift on the wing in a bank, then, must be greater than that in level flight. This means that the wing is experiencing a g-force greater than one, even though it is in a level turn, and not in a pull-up maneuver.

If the vertical component of lift is equal to the total lift times the cosine of the bank angle (which must equal the weight), the total lift, then, must be equal to the weight divided by this cosine. We can think of it as equal to the weight multiplied by a factor of one divided by the cosine of the angle:

$$L = \frac{1}{\cos\phi} \times W$$

From the section on maneuvering in chapter 5, we can see that this is the same as our definition of lift under accelerated conditions: where

$$L = n \times W$$

Thus, the term $1/\cos\phi$, is the same as the load factor, when bank is involved, and this fact enables us to establish a relation between bank angle and load factor or g-force:

$$n = \frac{1}{\cos\phi}$$

If we calculate n for various bank angles from 0 to 90 degrees, the results are as shown in Fig. 6-23. For example, at 60 degrees, the cosine is 0.5, and 1 divided by 0.5 is 2; thus, in a 60-degree bank, the airplane is experiencing a 2-g acceleration. At 90 degrees, the n goes to infinity, which means that an infinite amount of lift would be required, or flight at this condition is impossible. This is actually the case if only the wing were involved. In so-called *knife-edge* air show maneuvers, where the airplane is flown sideways, lift is generated by the side of the fuselage and vertical fin, as well as a vertical component of thrust when the nose is angled upward.

Chapter 5 also noted that the stall speed increased proportional to the square root of the load factor; thus, if higher bank angles generate higher load factors, stall speeds also increase. Figure 6-24 is a plot of stall speed factor versus bank angle. At 60 degrees, for example, the stall speed factor is 1.4. For example, if the level flight stall speed were 50 knots, the stall at 60 degrees would occur at 1.4 times 50, which is 70 knots.

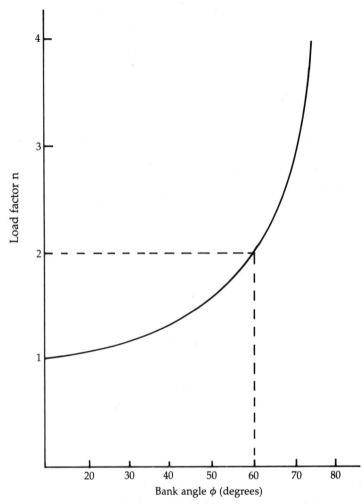

Fig. 6-23. Load factor versus bank angle.

The increase in stall speed and load factor with bank angle imposes a limit on turn performance. Figure 5-25 showed that the limit load factor for the airplane depicted was 3.8. Figure 6-23 indicates that the bank angle at 3.8 g's is approximately 75 degrees. Any further bank would exceed the structural design limits. The stall speed line intersects this upper limit load line at 117 knots; thus, the airplane cannot be slowed any more than this speed, or it would stall. Because the rate of turn was seen to be proportional to the angle of bank and inversely proportional to the velocity, this corner of the V-n envelope represents the conditions for maximum rate of turn. For this airplane, using the formula given above for turn rate, the maximum rate of turn for this airplane would be approximately 35 degrees per second.

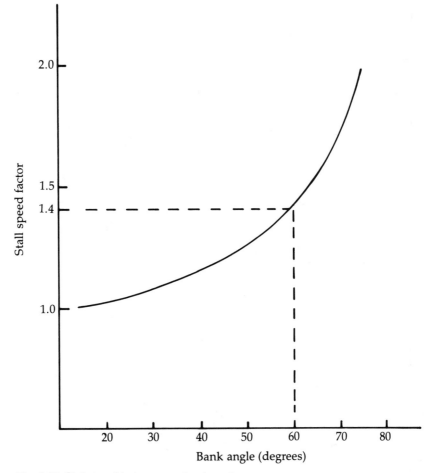

Fig. 6-24. Stall speed factor versus bank angle.

The smallest radius of turn also occurs at this same set of conditions because it was shown to be proportional to velocity squared and inversely proportional to bank angle. For this airplane, the smallest radius of turn would be 325 feet.

## Power limits on turn performance

The section on level flight performance in chapter 5 explains how drag determines the power required to fly in level flight; however, that definition for power required was for straight and level flight. When the airplane is banked into a turn, more power is required for level flight than a wings-level condition. Figure 6-25 shows power required curves for a typical lightplane at 45° and 60° of bank compared with the same airplane at 0° bank angle.

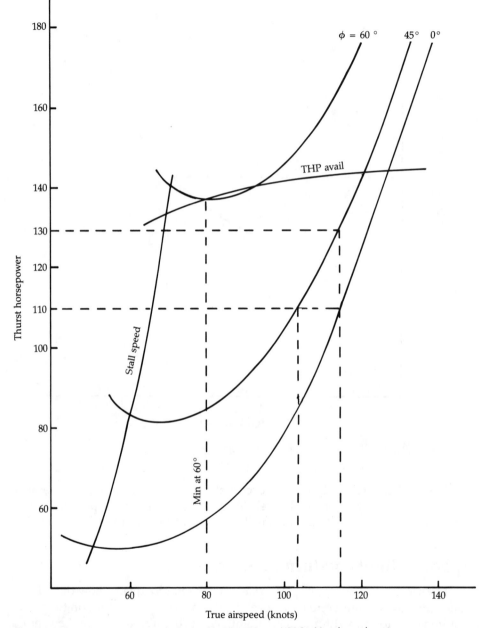

Fig. 6-25. Power curves for an airplane with 0°, 45°, and 60° of bank angle.

Notice that if the airplane were flying straight and level at 115 knots, it would require 110 thrust horsepower. Now, if the airplane were rolled into a 45-degree bank, 130 horsepower would be required to maintain level flight at the same airspeed. This is due to the fact that the increased lift required for the turning flight requires a higher angle of attack, which produces more induced drag. If, however, the airplane were slowed down to 105 knots, notice that the 110 horsepower would still be sufficient to maintain level flight in the 45° bank. Because both radius and rate of turn are improved at lower airspeeds, speed reduction rather than power increase is the common control input to accompany a bank.

However, at very steep bank angles, speed reduction alone might not be enough. Notice the power required curve for 60° of bank. A minimum of approximately 137 horsepower is required for this attitude without any altitude loss. For the example airplane, this is about as much power as there is available; therefore; the airspeed for such a bank angle is very critical. Level flight can only be maintained between 80 and 93 knots at this angle.

A further limitation on turn performance now becomes apparent. Notice that the stall speed, which we previously considered to be the limit on turn performance is 71 knots, but the minimum level flight speed due to power considerations is 80 knots; hence, this speed determines the minimum radius and maximum rate of turn, rather than stall speed. Jet fighter aircraft are also often limited in turn performance by thrust available. This fact becomes particularly evident at high altitudes, where the induced drag is high for even straight and level flight. Induced drag naturally increases substantially during a steep bank.

# 7

# High-Speed Flight

WE HAVE SEEN THAT ANYTHING MOVING THROUGH THE AIR affects the air pressure in its vicinity. This pressure change will also affect the density of the local air; however, at relatively low speeds such as 100 or 200 mph the density change is so slight that we can normally neglect it altogether. In such low-speed flight regimes, the air is *incompressible*, meaning that the density of the air flowing over the aircraft does not change from that of the surrounding *ambient* air.

This situation is not true of high-speed flight. In the late 1930s, when airplanes first reached speeds of 300 to 400 miles per hour, it became evident that significant density changes did occur in the airstream flowing around the airplanes. Such airflow was termed *compressible flow* because the air actually does compress. When it does, unusually large increases in drag can occur. As airplanes travelled faster and faster, they encountered drag considerably higher than predicted by the usual methods used to describe low-speed flight. Scientists theorized that as airplanes approached the speed of sound, the drag would go to infinity and could not be overcome no matter how much thrust could be produced. Indeed, this concept was held by many of our leading aeronautical experts well into the 1940s; hence, the term *sound barrier* originated.

Other experts contended that this idea was wrong and that flight at and above the speed of sound could be achieved with the correct configuration. Bullets, after all, had been travelling faster than the speed of sound for many years; the first airplane to fly faster than the speed of sound was based on the design of the Browning .50-caliber machine gun bullet. The Bell X-1 (Fig. 7-1) was first taken to supersonic speed by test pilot Chuck Yeager on October 14, 1947. Like the flight of the Wright brothers in 1903, this was no wild shot in the dark, but the culmination of extensive and painstaking research and testing spanning many years.

## THE SPEED OF SOUND

Undoubtedly, you have heard before of the terms *speed of sound*, *supersonic*, and *sound barrier*; however, you might wonder what in the world sound has to do

Fig. 7-1 The Bell X-1, first airplane to exceed the speed of sound in 1947.

with drag on an airplane. Actually, sound in itself has nothing to do with it. An airplane does not "hear" anything.

The significance of the speed of sound is that sound is due to pressure disturbances in the air. We hear sounds because of these pressure disturbances being transmitted to our ears in wavelike fashion through the air. This transmission takes place by a series of molecules bumping into adjacent molecules until finally those just next to our eardrums bump into them.

The pressure transmitted to the surrounding air by an aircraft flying through it is carried by the same mechanism, *molecular collision*. The pressure will thus be carried at the same rate as the speed of sound. If the airplane is flying slower than the speed of sound, the pressure disturbance can be transmitted ahead of it. This is the process, for example, by which upwash is created ahead of the wing. You could also hear an airplane approaching you if it were slower than the speed of sound; you would not hear a supersonic airplane approaching. Likewise, the air ahead of a supersonic airplane would have no warning of its approach and could not begin to flow in a pattern that would carry it smoothly over the aircraft surface until it actually touched it. The point is that sound is actually *pressure transmission* and the aerodynamic effects of pressure are carried at the same rate as that of sound.

You might wonder now what the speed of sound actually is in numbers. The answer is that it varies and the variation is caused by the temperature of the air. Temperature is actually a measure of the molecular energy of a body. If air (or a piece of metal, or anything else) is relatively hot, its molecules are moving approximately at a relatively high rate of speed. A hot piece of metal feels hot to the touch because its molecules are striking your fingers at a high rate of speed. Actually, all of the molecules do not move at the same rate, so that temperature is really a measure of the *mean* rate or *mean* amount of energy of the molecules. There is an old saying among scientists in the wintertime that "The mean molecular intrinsic energy of the atmosphere is equal to that of the cardium of a sorceress." This is just a complicated way of saying that it is as cold as a witch's heart.

Because molecules move about at a higher rate of speed in warmer air, they are able to bump into each other quicker, and thus transmit sound faster; therefore, the speed of sound depends on the temperature of air. At standard sea level temperature of 59°F, the speed of sound is approximately 761 mph. At −12°F, which is the standard temperature at 20,000 feet, the speed of sound is 707 mph, and at 40,000 feet, where the temperature is approximately −70°F, the speed of sound is only 660 mph. The actual physical relationship of the speed of sound at any altitude to that at sea level is the square root of the temperature ratio. That is, the speed of sound for any altitude is the speed of sound at sea level multiplied by the square root of the temperature at altitude divided by standard sea level temperature:

$$V_{sound} = V_{sound_{sl}} \sqrt{\frac{T}{T_{sl}}}$$

## MACH NUMBER

Because of the significance of the speed of sound in high-speed flight, flight velocities in this realm are usually measured as a ratio between the actual speed and the speed of sound. This method of measuring airspeed was developed by the Austrian physicist Ernst Mach. The value has thus become known as the *Mach number* and is given the symbol M:

$$M = \text{true airspeed/speed of sound}$$

Therefore, if an airplane is travelling at one half the speed of sound, we say that it is going Mach 0.5; a speed of Mach 1.5 would be 50 percent greater than the speed of sound. Of course, the speed of sound used in the ratio would have to be for the altitude (or temperature) at which the aircraft is flying; hence, Mach 0.5 at sea level would actually be faster than Mach 0.5 at 40,000 feet.

Since the advent of high-speed flight, we usually classify speed into distinct regimes, or ranges of speed. In the simplest breakdown, we could say that speed right at the speed of sound, or Mach 1, is *sonic*, anything slower is *subsonic*, and anything faster is *supersonic*. In summary:

| | | |
|---|---|---|
| M less than 1 | : | subsonic |
| M = 1 | : | sonic |
| M greater than 1 | : | supersonic |

Flight precisely at Mach 1, however, is a rather unstable situation. Only very slight disturbances will cause it to go either subsonic or supersonic. Furthermore, an airplane flying at exactly Mach 1 will have some flow over parts of it at lower speeds and over other parts at higher speeds. Consequently, flight in this area is not purely subsonic, sonic, or supersonic. The term *transonic* was thus adopted for flight in the vicinity of Mach 1. Also, it was found that airflow also has distinct characteristics at very high Mach numbers. Flight at such speeds was termed *hypersonic*; therefore, a more complete and practical breakdown of speed regimes was developed:

| | | |
|---|---|---|
| M less than 0.8 | : | subsonic |
| M = 0.8 to 1.2 | : | transonic |
| M = 1.2 to 5.0 | : | supersonic |
| M greater than 5.0 | : | hypersonic |

You might see some variation in these definitions, particularly in the transonic range. Some sources define transonic as speeds between Mach 0.7 and 1.3, but it is generally agreed to include in the transonic area some percentage between 20 and 30 both above and below sonic velocity.

## SHOCK WAVES

In order to understand how high speed affects the compressibility of the air, let us consider an example of wave motion. Suppose we drop pebbles into a pool of water and observe the waves resulting from these disturbances. If we drop all pebbles at the same spot, but spaced at equal time intervals, the waves would all spread out from a single point as shown in Fig. 7-2A. The outermost wave would be from the first pebble, the next wave from the second, and so on.

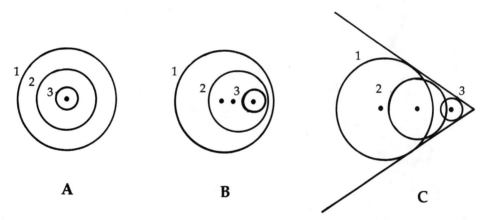

Fig. 7-2. A sequence of pressure disturbances at a stationary point causing waves 1, 2, and 3 to emanate from that point (A), pressure disturbances moving to the right, but at a speed slower than the wave movement (B), and moving at a speed greater than the wave movement causing a wave front (C).

Now suppose that we drop the pebbles at the same time interval but slowly move to the right as we drop them. The pattern would now appear as in Fig. 7-2B. Notice that the waves are moving faster than our forward speed. Finally, we proceed at a rate faster than the waves are able to move. Each pebble hits outside of the wave made by the previous pebble. Just as we drop the fourth pebble, the pattern looks like that of Fig. 7-2C. The waves pile up at the forefront of our motion and tend to reinforce each other.

Now assume that instead of pebbles in water these are pressure disturbances in the air (noise emissions, if you prefer). As long as the pressure waves travel faster than the source, we have disturbance ahead of the source and a subsonic flow condition; however, when the source travels faster than the wave progression, we have a supersonic condition and the waves pile up on each other. A line drawn tangent to the wave fronts in Fig. 7-2C shows the area of this reinforcement and results in *Mach wave*.

Mach waves result from minute, or point, disturbances. If we placed a larger object such as a wedge in a supersonic flow, a similar pattern would result but it would create a stronger disturbance known as a *shock wave*. Figure 7-3 shows a photograph of the shock wave on such a wedge in a supersonic wind tunnel. The shock wave is made visible by an apparatus making use of the deflected light rays shown through the denser air in the wave. Such shock waves are actually referred to as *oblique* shock waves because they are oblique, or at an angle, to the direction of flow.

Fig. 7-3. Schlieren photograph of a wedge in a supersonic wind tunnel showing the shock wave formation.

There is a drastic change of pressure across the shock wave because there is no change at all ahead of it and all of the pressure from behind has reached this front. The increase in pressure is sufficient to also cause an increase in the density; therefore, we say that there is a *compression* in the air at this point.

# CRITICAL MACH NUMBER

Recall from the discussion of lift that a gradual pressure decrease occurs along the top surface of an airfoil from the leading edge back to the area of maximum thickness. This pressure decrease is caused by an *increase* in the air velocity over the surface; therefore, the local velocity in this region of the airfoil is greater than the velocity of the free airstream. If the velocity is greater, then the Mach number must also be greater.

It follows, then, that some point on the airfoil will reach Mach 1 before the free airstream velocity is actually Mach 1. The value of the free airstream Mach number (that is, the forward velocity of the airplane) that causes the flow to just reach Mach 1 somewhere on the airfoil is called the *critical Mach number*. This is best explained by an illustration.

Figure 7-4A shows an airfoil in a free stream flow of Mach 0.7. The velocity increases over the forward part of the airfoil, but only reaches a maximum of 0.82; hence, the flow is all subsonic. In Fig. 7-4B, the velocity of the airplane is increased to Mach 0.85, causing the velocity to reach Mach 1 at the maximum point. The critical Mach number of this airplane with this airfoil is therefore considered to be Mach 0.85. This example also illustrates the varying Mach numbers over different parts of the airplane in the region of Mach 1 and the reason for transonic speeds. In this particular case, the transonic regime for this aircraft would begin at Mach 0.85 because slower than that is all subsonic flow. This of course assumes that no faster flow is on any other part of the airplane. Usually it is the wing that reaches sonic velocity first, in which case the wing dictates the critical Mach number.

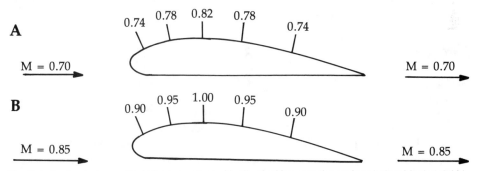

Fig. 7-4. A wing below critical Mach number with all velocities over the surface below Mach 1.0 (A), and a wing at critical Mach number with velocity just reaching Mach 1.0 on the top surface (B).

# AIRFOILS IN TRANSONIC FLIGHT

When the airplane exceeds the critical Mach number, a region of supersonic flow will exist on the airfoil. At the trailing edge, the flow must again assume the Mach number of the free airstream (the forward speed of the airplane). If the airplane is traveling below Mach 1, this means that the flow must slow down to subsonic as it approaches the trailing edge.

Figure 7-5A shows such a situation for an airplane traveling at Mach 0.87, which is slightly faster than its critical Mach number. Now, remember that pressure disturbances travel in all directions, just like noise or the pebbles in the pond. In the subsonic region, these disturbances can move forward; in the supersonic area, however, they cannot go forward of the disturbance. The barrier between these two regions becomes very distinct and leads to the formation of a shock wave. It will be located at the point where the flow has slowed to just Mach 1, designated by point 3 in the figure. The flow will also be Mach 1 at point 1; this is where the supersonic region begins. It will build up to a higher value (such as 1.05) at point 2 and then slow to Mach 1 at point 3. The important fact is that a shock will occur at this point where the flow changes from supersonic back to subsonic. Such shock waves are normal (or perpendicular) to the flow at that point and are referred to as *normal shock waves*.

If the speed of the airplane is increased, say to Mach 0.9, the shock wave will move farther back on the airfoil, with a resulting larger area of supersonic flow. By this time the flow has probably also speeded up enough to establish greater-than-sonic velocity on the lower surface. This results in a shock wave occurring on the *lower* surface as well, and the picture appears as in Fig. 7-5B.

As the speed is increased even closer to Mach 1, both shock waves will move to the trailing edge and almost the entire airfoil will be immersed in supersonic flow, as shown in Fig. 7-5C. The shock waves now actually become oblique because the flow turns from the direction of the airfoil surface to the free airstream direction as it passes through them.

If the speed were to exceed Mach 1, another shock wave would appear ahead of the airfoil, as shown in Fig. 7-5D. This sort of wave is known as a *bow wave*. Behind it there is a small region of subsonic flow surrounding the leading edge, but everywhere else the flow is supersonic.

## WAVE DRAG

When shock waves exist, they cause a significant increase in drag. Recall from the discussion of wakes in chapter 3 that separation can be caused by an adverse pressure gradient, which is a term for pressure that intensifies in the direction of the airflow. It arises, naturally, from decreasing velocity due to decreasing thickness in the aft part of the airfoil. The shock wave gives the same effect, but even more so. Instead of a gradual increase in pressure, there is a very sudden increase in pressure across the shock wave. This in itself gives rise to an increase in drag, because it takes kinetic energy out of the air.

In addition, when the shock wave is strong enough it can cause a rapid separation of the flow with a resulting wake formed behind it. Both upper and lower surface shock waves can cause such separation, as shown in Fig. 7-6. The combined effect of the energy loss across and the wake formed behind the shock wave is called *wave drag*.

At speeds very slightly above the critical Mach number, wave drag is due mostly to the energy dissipation and is rather small; however, as separation starts to occur, the drag increase becomes quite significant. At some point drag's

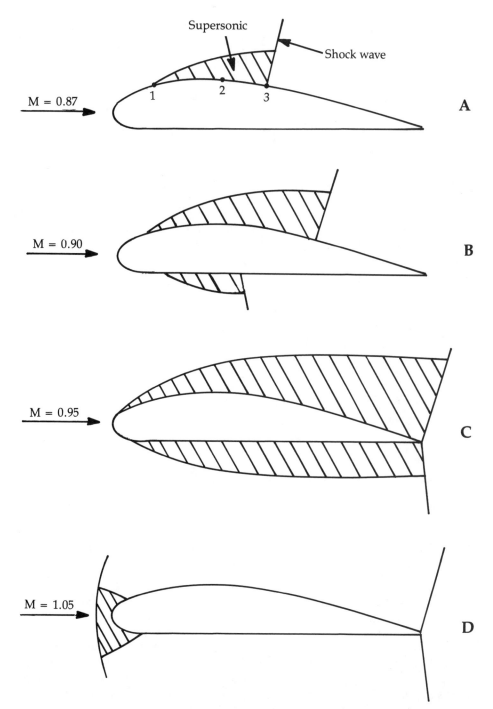

Fig. 7-5. Progressive shock wave formation on an airfoil with supersonic flow only on the top surface (A), beginning on the bottom surface (B), moving to the trailing edge (C), and finally, the entire airfoil at greater than Mach 1.0 (D).

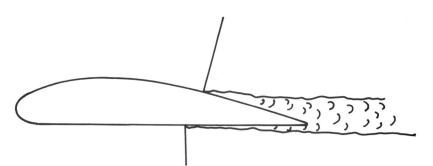

Fig. 7-6. Shock waves induced separated flow or wake formation.

increase with increasing Mach number becomes quite steep. This point is known as the *critical drag Mach number*, or sometimes the *drag divergence Mach number*. Figure 7-7 shows the variation in airfoil drag coefficient with Mach number.

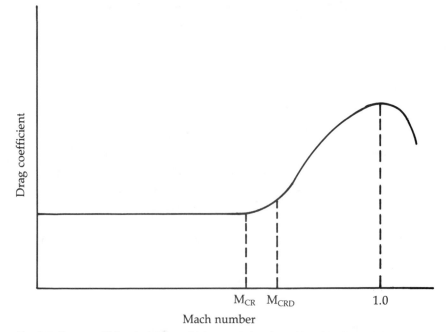

Fig. 7-7. Drag coefficient behavior at high speed showing critical Mach number and critical drag Mach number.

The drag coefficient is essentially constant below the critical Mach number. It begins to increase slightly directly above it. At the critical drag Mach number, it begins to shoot up dramatically. It is no wonder that early scientists, looking at test results in this region, speculated that the drag must build up to infinity at Mach 1; however, when research capability was developed to actually test at Mach 1 and above, they learned that the drag actually peaked at Mach 1 and then dropped off, as shown.

Getting to, or close to, Mach 1 was possible, it just took a tremendous amount of thrust. The jet engine made that thrust possible at reasonably low engine weight. It also eliminated the propeller, which of course involves an airfoil, and encounters the same sort of high-speed problems that wings do. In addition, the prop has added speed from its rotation, so that shock waves, with their increased drag effects, are encountered at relatively low forward speeds. The jet was the ideal powerplant for high-speed flight. At first such flight was achieved just from the brute force of high-thrust engines, such as used by the military and in early airliners. Eventually, engineers found ways of keeping airplanes below their critical drag Mach numbers and speeds of 500 to 600 mph were achieved along with fairly economical operation. These developments made possible the introduction of the jet to general aviation. One of the first such aircraft, and certainly the most famous, was Bill Lear's original Lear Jet, shown in Fig. 7-8.

Fig. 7-8. The original Lear Jet Model 23.

Nevertheless, flight in the transonic region is a tricky business, and much research is still going on in this area. It should be obvious from Fig. 7-7 that flight in the vicinity of Mach 1 is not very practical. Drag is reduced and aerodynamic behavior is simpler if you go significantly beyond Mach 1—that is, faster than the transonic region. On the other side, it is wise to stay well below the peak in the drag curve. Of course, subsonic flight can be made faster if the critical Mach number and critical drag Mach number can be made higher. This is the object of much research in the area of high subsonic flight.

## SWEPT WINGS

One of the first breakthroughs that allowed for high critical Mach numbers was the idea of wing sweep. If a wing is swept aft, only a component of the velocity of the air will flow over it chordwise. Another component will flow spanwise along the wing. This allows the airplane to fly at a higher Mach number, while the wing's airfoil only "sees" a portion of this speed. To illustrate, consider the wing in Fig. 7-9. Suppose that the airfoil in this wing has a critical Mach number of 0.75. If the wing were straight (unswept), then 0.75 would also be the critical Mach number of the overall airplane.

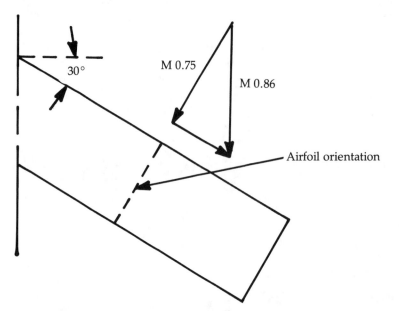

Fig. 7-9. Wing sweep reduces the effective Mach number along the chord.

Now sweep the wing 30°. Only a component of the forward velocity will hit the wing chordwise. Actually, this component is equal to the cosine of 30° (0.866) times the free airstream Mach number; thus, if the airplane were going Mach 0.86, the component going over the airfoil would be only at Mach 0.75 and not exceeding the critical Mach number. The airplane's critical Mach number has been effectively raised to 0.86. This is really an ideal situation, and in actuality, the critical Mach number is not increased quite this much.

Unfortunately, the wing's low-speed performance is degraded by sweep. Remember that a significant part of the air velocity is now flowing spanwise and not contributing to lift. This will raise the stall speed, and the resulting takeoff and landing distance over an equivalent straight wing; however, much of this performance can be saved with the use of extensive flaps or other high-lift devices. Spanwise flow will also cause the tips to stall first, which could cause a loss of aileron effectiveness before the entire wing is stalled. Again, this detriment can be overcome by use of flow fences running chordwise on the wing. We never said that high speed flight was not complicated and expensive.

Swept wings do have some other advantages. Chapter 6 pointed out that sweep improves the lateral stability of an airplane, giving the same effect as dihedral. For supersonic aircraft, sweeping the wing to a greater angle than the Mach wave formed by the wing root can place the entire wing in subsonic flow, thereby reducing wave drag. This situation is shown in Fig. 7-10.

Wings can also be swept forward with most of the same results. Spanwise flow is even in a desirable direction with forward sweep and lift properties are thus improved. Until recently, forward sweep yielded serious structural problems and often generated flutter. New composite structures are overcoming those problems and seem to hold great promise for forward-swept configurations in the near future.

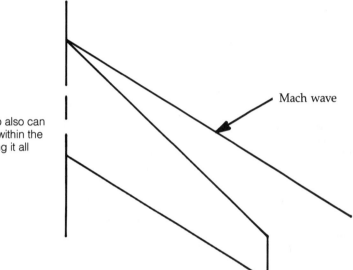

Fig. 7-10. Wing sweep also can place the entire wing within the Mach wave, thus giving it all subsonic flow.

Mach wave

## HIGH-SPEED AIRFOILS

The ideal airfoil for supersonic flight is one with a sharp leading edge. Figure 7-11 shows a biconvex airfoil, a shape typical of supersonic airfoils. Note that the sharp leading edge allows for the bow wave to actually attach itself to the leading edge. It then becomes an oblique shock wave and the entire airfoil encounters supersonic flow.

Such airfoils have very poor low-speed characteristics and are only used on military fighters and missiles. Trainers, airliners, and business jets require a more conventionally shaped airfoil with a somewhat rounded leading edge. In order to enable high critical Mach numbers, these airfoils have to be somewhat different compared to those used for very low-speed operation. For one thing, such airfoils need to be as thin as possible. This characteristic prevents the airflow over the airfoil from being speeded up very much from the free airstream velocity. Another desirable trait is to have the point of maximum thickness very far back on the airfoil. This creates a favorable pressure gradient and tends to

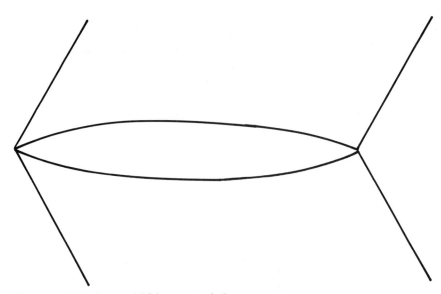

Fig. 7-11. A circular arc airfoil in supersonic flow.

reduce drag at high speeds as well as it does at lower speeds. The 6-series, or so-called *laminar flow* airfoil discussed in chapter 2, was the ideal choice at the time high subsonic flight became possible. Lower thickness ratio laminar flow airfoils are still in use today on many airliners and business jets.

Science marches on and NASA researchers developed an airfoil with a critical Mach number very close to Mach 1. For this reason, the airfoil was termed a *supercritical* airfoil. This extremely efficient high subsonic airfoil was developed with the aid of advanced computer techniques. The development is credited largely to an already famous NASA engineer, Dr. Richard Whitcomb. Dr. Whitcomb also applied the same technology used in supercritical airfoil development to develop super-efficient low-speed airfoils. The GAW-1 and GAW-2 discussed in chapter 2 were the results. These airfoils resemble the supercritical airfoil in their general shape, and are sometimes mistakenly referred to as "supercritical."

The supercritical airfoil is shown in Fig. 7-12. The maximum thickness of this airfoil is very far back toward the trailing edge. In addition, the upper surface has very shallow curvature. This prevents the pressure from having a large peak over the top surface. Instead, it is fairly flat over much of the airfoil. The flow is only slightly supersonic, but covering a large portion of the upper surface, at critical Mach number. The flow is then very gradually decelerated near the trailing edge to subsonic speed to discourage the formation of a shock wave. This results in low drag even at speeds close to Mach 1. Supercritical airfoils are seeing increased use in newer airliner and business jet designs.

## CONTROL PROBLEMS

Figure 7-6 demonstrated how the shock waves can cause separation to occur on the airfoil surfaces and form small wakes behind them. This is the same sort of

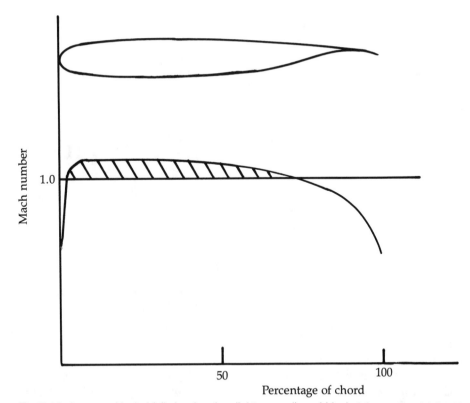

Fig. 7-12. A supercritical airfoil showing the slight exceeding of Mach 1.0 over the airfoil and the gradual transition to subsonic flow with little or no shock wave formation.

phenomenon as a stalled condition. Now suppose that there is a control surface, such as an aileron, at the trailing edge. Obviously it would be very ineffective in this stalled region, particularly at the low deflection angles typical of such surfaces. To overcome such problems, *vortex generators* were employed.

Vortex generators are shown in Fig. 7-13. They are essentially small winglike surfaces that project vertically into the airstream. A vortex is formed at the tip of each protrusion, as at the tip of an ordinary wing. The vortices from a number of these generators placed ahead of the control surface adds energy to the air in the boundary layer from outside the boundary layer. This additional energy enables the boundary layer to overcome the adverse pressure gradient of the shockwave and prevents separation. In addition, the vortex generators tend to break up the shock waves. Both of these actions tend to delay separation and give greater control effectiveness. They also reduce the wave drag. This fact makes up for the additional drag created by the vortex generators.

Another problem that can occur in supersonic flight has to do with conventional control surfaces with fixed and movable portions, such as a stabilizer-elevator combination. At supersonic speeds a shock wave will form at the juncture of the stabilizer and elevator, as shown in Fig. 7-14A. The shock wave prevents any influence of the elevator on the stabilizer ahead of it. Remember that nor-

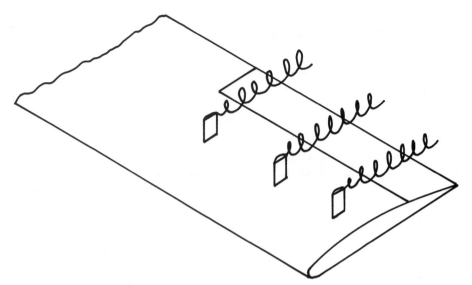

Fig. 7-13. Vortex generators to pull high-energy air over a control surface at high speeds.

mally an elevator acts like a flap and gives increased lift to the entire horizontal tail surface. The amount of pitch control is thus greatly reduced in the supersonic condition.

Ingenuity again won out when the all-movable tail surface came into being. Such a surface eliminates the junction and the resulting shock wave, and allows for full surface effectiveness, even at supersonic speeds, shown in Fig. 7-14B. This problem first came to light about the time of the Korean War when fighters were first reaching or slightly exceeding the speed of sound. All-movable surfaces on American F-86 Sabre Jets gave them an edge over the faster and more powerful Russian MiG-15s, which had conventional controls. Attention to technical details such as this, along with superior training methods, enabled us to hold our own in a war in which we were, in general, technologically behind the enemy.

Research in all regimes of high-speed aerodynamics continues to yield developments that lead to improved airplane designs: fighters capable of flight faster than Mach 3, commercial transports such as the Concorde that are capable of Mach 2, and light jets such as the Cessna Citation that are as easy to fly as many prop-driven light airplanes (Fig. 7-15).

## AREA RULE

When supersonic fighters first came onto the scene in the early 1950s, it was found that considerable drag was encountered in the region of Mach 1, requiring a hefty amount of thrust to overcome it. Much of this drag was wave drag, the additional pressure drag created when shock waves form on the aircraft. Richard Whitcomb provided a big breakthrough in the technology of this phenomenon with what has become known as the *area rule*.

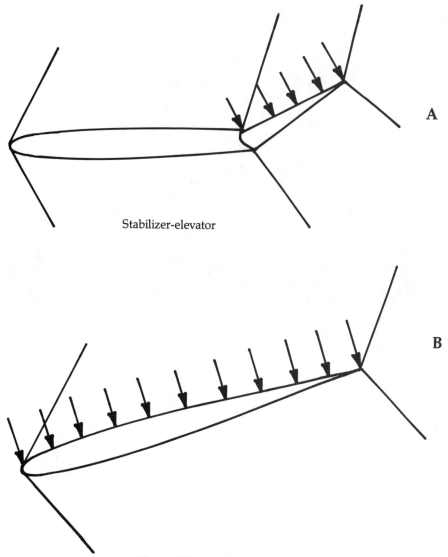

Stabilizer-elevator

All-movable stabilator

Fig. 7-14. Stabilizer-elevator combination in supersonic flow, limiting control force to elevator alone (A), and all-movable stabilator in same flow (B).

Whitcomb experimented with a number of wing-body shapes in a high-speed wind tunnel, and came up with a very significant finding. He found that for low-aspect-ratio wings the drag rise in the transonic region was primarily dependent on the progression of cross-sectional area of the entire aircraft from nose to tail, regardless of the exact shape of the cross sections. To illustrate, he found that, when swept wings were added to the cylindrical fuselage shown in Fig. 7-16(A) to form the configuration of Fig. 7-16(B), the parasite drag at Mach 1

Fig. 7-15. Cessna Citation I.

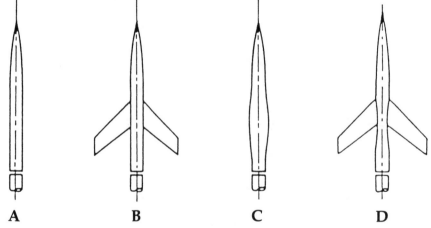

A            B            C            D

Fig. 7-16. Models used for Whitcomb area rule tests: cylindrical fuselage (A), fuselage with swept wing (B), bulged fuselage (C), and reduced area fuselage (D).

doubled. The wave drag accounted for most of this rise by increasing approximately four times. This result appeared to be due to the sharp increase in total cross-sectional area at the wing root. The same drag rise occurred if he simply added a bulge in the cylinder equivalent to the volume added by the wing, as seen in Fig. 7-16(C). This finding supported the theory that the drag depended more on the amount of cross-sectional area rather than the shape.

In order to smooth out the sharp increase in area caused by the wing, he reduced the area of the fuselage in the region of the wing, as shown in Fig. 7-16(D). The result was a dramatic reduction in total drag, with only very slight increase in wave drag. The drag was just about the same as fuselage alone if the fuselage were reduced in cross-sectional area equivalent to the additional contribution to overall cross-sectional area added by the wing. It turns out that there is an ideal shape for a body with given length and diameter to minimize the wave drag. Such shape is known as a Sears-Haack body (after the discoverers of this principle; not the store that sells them) shown in Fig. 7-17. The closer the overall cross-sectional area distribution comes to this shape, the lower the wave drag will be.

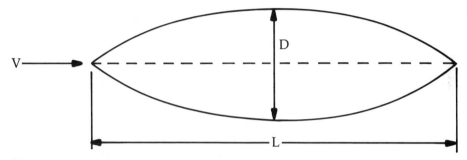

Fig. 7-17. Sears-Haack body.

The area rule found immediate application in the next generation of supersonic fighters. One of the first to employ this concept was the Convair F-102 (Fig. 7-18). The fuselage was reshaped to the configuration shown in Fig. 7-19, and designated the F-102A. This configuration became more commonly referred to as the *coke-bottle* fuselage. Indeed, it became such a trendy shape that it was copied by a famous American sports car, which, of course did not really benefit from wave drag reduction in the Mach 1 speed range. The original coke-bottle shape was pretty much limited to one- or two-place fighters; the reduction in fuselage capacity was considered to be prohibitive for use in supersonic transport design; however, the idea is so effective in drag reduction at high speeds that it is being considered in plans for the next generation of supersonic transports.

## HYPERSONIC FLIGHT

At the speed of sound, Mach 1, some very distinct changes occur in the airflow. Shock waves form, and the pressure and density distributions resulting cause all kinds of changes to the lift, drag, and other forces. The effect on drag coefficient, for example, can be seen in Fig.7-7; hence, there is a very definite boundary between subsonic and supersonic flow. Once we get through the transonic range (approximately 20 percent faster and 20 percent slower than Mach 1), the aerodynamic effects again pretty much stabilize up to approximately Mach 3 or so; however, when flying even faster, a number of physical aspects of airflow again begin to change significantly.

Fig. 7-18. Convair (General Dynamics) F-102 fighter.

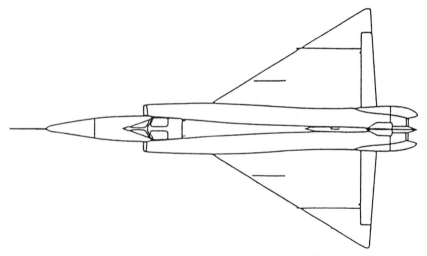

Fig. 7-19. Diagram of Convair F-102A fighter showing area-ruled fuselage.

Some of these effects will occur in the region of Mach 3, while other effects are not of much consequence until approximately Mach 7. Above this speed, there are considerable differences from flow at Mach 2, for instance; therefore, this region is referred to as the *hypersonic* regime of flow. Because there is no distinct boundary where any bells sound or lights flash, the designation of a limit Mach number for hypersonic flight is somewhat arbitrary. It is usually considered to be Mach 5 because that is a good average of the Mach numbers where the various hypersonic effects take place.

## Aerodynamic effects

One of the differences in the flow at very high Mach numbers is the squeezing of the shock layer. The shock wave shown in Fig. 7-3 resulted from a flow at approximately Mach 2.5. The angle that the shock wave makes with the flow, or centerline of the wedge, depends on the Mach number. As the Mach number gets higher, this angle gets smaller. In the hypersonic region the angle is very small, as shown in Fig. 7-20. The streamlines passing through the shock are bent, and pass between the shock wave and the body surface. This region is referred to as the *shock layer*. Because it is very small for this high speed flow, the shock layer can interact with the boundary layer (described in chapter 3).

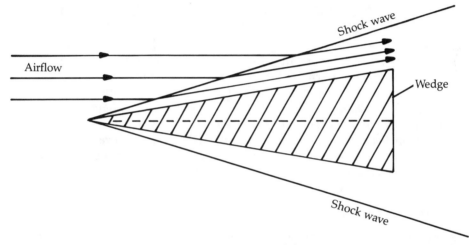

Fig. 7-20. Shock wave at hypersonic speed showing shock layer.

Such interactions can result in considerable effect on drag. They also can cause unusual thermal reactions and contribute considerably to aerodynamic heating. The actual details are very complex, but they are much different from the effects at supersonic speeds below the Mach 5 range. Without going into the exact details, it should be apparent that we are dealing here with a different set of physical conditions; hence, vehicles designed for hypersonic flight are significantly different from what we normally term "supersonic aircraft," which usually implies flight around Mach 2 or so.

The shock wave pattern at these Mach numbers also requires a highly swept, wedge-shaped wing. For the high end of the envisioned hypersonic speed range, the wing will have to be merged into the fuselage so much that it will become an integrated shape, with the actual wing area hardly distinguishable.

## Aircraft concepts

Recently, there has been much attention given to the design of an aircraft for operation in this speed range. It has been referred to at various times and by various agencies in the United States as the *hypersonic transport* (HST), the *Orient Express* (because it was envisioned as being used primarily for flights between the United States and the Orient), and the *aerospace plane*. This last designation comes from the fact that it is envisioned as a hybrid aircraft and spacecraft. This means that it would take off and land like an airplane, but fly high enough and fast enough that it could go into orbit on its own momentum for the cruise segment of flight.

Such a vehicle would be a step beyond the space shuttle because it is envisioned as using air-breathing engines, achieving all acceleration within the atmosphere. Furthermore, it would be a single-stage propulsion system; no booster rockets or any such device would be used for initial acceleration, and then drop off.

Planners for the commercial transport *Orient Express* version are looking realistically at speeds of Mach 5 to 8. Figure 7-21 shows the Mach 5 HST proposed by Northwest Airlines, which could go from New York to Tokyo in fewer than three hours. The X-15 research aircraft actually reached Mach 6.7 in the 1960s; however, studies under a government plan for a hypersonic research vehicle, which is being called the *National Aerospace Plane* (NASP), show that speeds up to Mach 25 are actually feasible. The space shuttle reenters the atmosphere at this speed, and the Apollo vehicles returning from the moon reached even higher Mach numbers; however, these vehicles are accelerated by huge, powerful, multistage rockets. To power a vehicle to such velocity with a single-stage, air-breathing engine is a real challenge.

## Propulsion systems

The only airbreathing engine developed so far that is capable of speeds much above Mach 3 is the ramjet, described in chapter 4. This engine compresses the air by the sheer ram effect as it enters the intake nozzle; however, conventional ramjets diffuse the air to subsonic speeds before fuel is injected and ignited. This makes the combustion process easier to handle, but the temperature in the combustion chamber gets very high. At the very high hypersonic speeds, the temperature would rise excessively from a structural material standpoint and the fuel would decompose before it could burn; thus, a new concept in ramjet engines is envisioned for the aerospace plane. This is one in which the flow would be all supersonic, even in the combustion chamber, which would tend to keep the temperature within manageable range. Such engines are referred to as *supersonic combustion ramjets*, or *SCRAMjets*. One of the major problems with this concept is ignition in supersonic flow; it has been described as being like trying to light a match in a tornado.

The SCRAMjet also demands an extensive, critically shaped inlet and exit nozzle arrangement; therefore, preliminary designs for hypersonic aircraft include significant consideration of the engine in the overall configuration. Many designs use the entire undersurface of the aircraft as part of the engine inlet and exhaust. Figure 7-22 shows the extensive exhaust nozzle of the launch vehicle of a two-stage HST as proposed by Boeing in earlier studies of potential hypersonic

Fig. 7-21. Hypersonic transport as perceived by Northwest Airlines.

designs. In a way, you could say that the aerospace plane will be an airplane designed around an engine.

Another problem with the propulsion system for hypersonic vehicles is the fact that even conventional ramjets are not practical at low speeds. They require high velocity for the ram compression. Indeed, they cannot operate at all if the

Fig. 7-22. Early concept of two-stage HST showing extensive engine exhaust nozzle area.

aircraft is not in motion; therefore, some other propulsion concept must be employed to initially accelerate the aircraft on takeoff and climb. It is not in the best interest of efficiency to have an auxiliary powerplant on board, which would only be used for low-speed flight, and then be carried along as additional weight in cruise. The solution seems to be a hybrid engine that could function like a conventional turbojet at low speeds, and then have the turbine bypassed by flow into a ramjet section at faster velocities. The term *turboramjet* has been given to this concept. Perhaps, for this particular application, it will be called a *turboSCRAMjet*.

# 8

# Design

UP TO THIS POINT we have been examining how the aerodynamic process works. It is just the natural laws of physics applied to airflow over an aircraft. Now let us look at how aerodynamics can be used. The object of aerodynamic research is, of course, to provide the technology to build better aircraft. Incorporating that technology into an efficient configuration is the process known as *design*.

Design is not a well-defined, step-by-step process. There are many different approaches to it. It is a science and an art. Like flying, it cannot be taught strictly from books; it must be practiced.

Aerodynamics, of course, is only one aspect that must be considered in design. Structures, propulsion, stability, and control are other branches of aeronautical science that must be taken into account. Modern aircraft design involves a large team of personnel with specialists from all of these areas and more; however, aerodynamic considerations form the basis for a design and this aspect serves as the starting point in the design process.

## DESIGN SPECIFICATIONS

Before a design can actually be undertaken, a set of specifications or requirements that the design should meet must be developed:

- Payload (passengers, cargo, or weapons)
- Range (or endurance)
- Cruising speed (or top speed)
- Takeoff and landing distance (field length)
- Ceiling

The aircraft company designing a new airplane will usually establish requirements based on market surveys and an understanding of the current available technology. For example, a company might decide that a feasible and competitive new airplane should have the following performance:

- Payload: Pilot plus five passengers.
- Range: 1,000 nautical miles.
- Cruising speed: 200 knots.
- Field length requirement: Minimum of 2,000 ft.

These specifications would be passed on to an engineering department that would then proceed with a design aimed at achieving this performance. In some cases it is not possible to meet all of the specifications; in others it might be possible to even exceed them. The latter situation often pleases bosses and wins raises.

In the case of military aircraft, the particular service will establish the requirements based on their particular needs. Sometimes (though rarely) there are even aircraft requirements established jointly among the various services. These requirements are passed on to qualified aircraft companies, who then prepare proposals to bid on the design and construction of airplanes to meet the requirements. Airline companies also sometimes follow this procedure; however, in modern times, the airlines and the aircraft companies are very much aware of the other's situations and new airline specifications usually arise from joint discussions and agreements.

In addition to performance requirements, current economic conditions have forced companies to include a number of requirements in this area. Until approximately the decade of the 1960s, cost was usually a secondary consideration to performance. This was especially true with military aircraft; however, by that time aircraft costs began to rise very sharply. Cost became a big factor and was often included in the primary specifications for an aircraft.

In the 1970s, the decisions of a few Middle Eastern oil barons forced a new item to be included in the specifications: fuel consumption. Unions, inflation, dwindling resources, and a number of other factors in a constantly changing world have greatly increased the costs of maintaining aircraft. Designers were thus forced to take into consideration another characteristic of an aircraft, namely, how easy it will be to fix. This characteristic is termed *maintainability*. Both the high cost of maintenance and the high replacement cost make it very desirable to have an airplane last a long time before it needs any significant maintenance, or replacement; hence, another term appears on the specification horizon that is called *reliability*. Economic requirements thus often prompt the addition of several, or all, of the following specifications to the list of performance items:

- Cost
- Fuel consumption
- Maintainability
- Reliability

## AIRWORTHINESS REQUIREMENTS

In addition to meeting the specifications, the designer must be concerned with following certain regulations and maintaining certain standards, established primarily in the interest of safety. Such standards are usually set by the federal government.

In the United States, the Federal Aviation Administration (FAA) is responsible for safety regulation of all civil aircraft. (The particular military service, of course, regulates such matters applying to its aircraft.) Federal Air Regulations Part 23 and Part 25 cover the requirements for the design of airplanes. Part 23 applies to light airplanes that by definition are airplanes of 12,500 pounds gross weight or less. Any airplane of higher weight must be designed to the standards included in Part 25. The weight used for this purpose is the certificated takeoff gross weight.

Part 25 is officially titled "Airworthiness Standards: Transport Category Airplanes." Part 23, which is applicable to most general aviation airplanes, is called "Airworthiness Standards: Normal, Utility, and Acrobatic Category." Light airplanes are certified in one or more of these categories. Most airplanes are certified in Normal category. Utility category allows for the performance of limited maneuvers, as usually required in certain types of flight training. Acrobatic category certification permits wringing out the airplane in full acrobatic maneuvers. Obviously, the requirements on design are more stringent for the Utility and Acrobatic categories. The most notable difference is probably that Utility and Acrobatic category aircraft must be designed to withstand higher inertia loads, usually known as *g loads*. As such, the structure must withstand higher stresses and consequently ends up being heavier than that of Normal category aircraft.

Part 23 is the bible for lightplane designers. Sometimes its requirements prevent the attainment of the desired specifications. Nevertheless, the requirements must be adhered to. A safe 150-knot airplane is much better than one that goes 170 and sheds its wings in the process.

# DESIGN PHASES

While most aircraft design is performed by a large team of experts from many different areas, everyone does not begin on his particular area simultaneously. The landing gear group does not begin designing the nose gear the minute the specifications are received. Nor does the fuselage group start work on the doors. Such things have to follow earlier decisions on configuration. A time sequence of events must occur to lead up to the final design configuration.

The sequence of events occurs in rather distinct phases. The entire design process is usually divided into three phases:

- Conceptual design
- Preliminary design
- Detail design

*Conceptual design* is the first step. It involves just what the name implies: a very basic, general concept of what the airplane will look like. A whole lot of sketching goes on in this stage. Here basic characteristics are considered, such as: will it be a jet or prop-driven, will it be single or multiengine, will it be high- or low-wing, and will it have fixed gear or retractable?

To an extent, the specifications will decide some of these questions. Obviously, if the cruising speed is to be 500 knots, it must be a jet. Or if it is to carry 200 passengers, it will probably end up having more than one engine. A two-place general aviation trainer, on the other hand, will dictate a fixed-gear, single-engine, prop-driven configuration. Other characteristics might be more arbitrary and are debated and decided on usually from the judgment of experienced designers.

Once a basic concept is arrived at, the design goes into the preliminary phase. *Preliminary design* is sometimes considered as being synonymous with aerodynamic design, and detail design is often thought of as structural design. To some extent this is true, but not entirely. Certain structural items must be considered in the preliminary phase, and some minor parts are shaped for aerodynamic reasons in the detail phase; however, preliminary design involves, largely, the consideration of aerodynamics in arriving at an overall configuration—the external shape of the airplane.

*Detail design* is the final stage, and consists mostly of the design of the supporting structure. Such structure must withstand the stresses imposed by the aerodynamic loads on the airplane and must also fit within the shape defined in the preliminary stage. It should be obvious that work in this stage often requires some modification to preliminary design decisions. It might be impossible, for example, to design a spar strong enough and yet light enough to carry the wing loads within the thickness of the wing specified in the preliminary design. This might require the preliminary design group to modify the wing to use a thicker airfoil. Likewise, analysis of the design in the preliminary phase might prove a change in basic concept to be the wisest way in which to proceed. In design, almost everything is tentative until the shop actually starts to cut metal.

Because aerodynamics is the subject of this book, discussions will be limited to conceptual and preliminary design. Structural design, which dominates the detail design phase, is a broad area in itself and is beyond the scope of this book.

# THE DESIGN PROCESS

The newcomer to aircraft design usually experiences a common emotion: frustration. No matter how he tries to improve the performance of the airplane in one area, it ends up detracting from the performance in others. For example, high cruise speed demands low drag, so the designer chooses a small wing area to keep drag down; however, he finds that this wing stalls at a very high speed, resulting in long takeoff and landing distances. He can remedy this by adding more power for takeoff and extending full flaps for landing; however, both moves add weight to the airplane, which reduces speed, as well as detracting from nearly all other performance—to say nothing of added cost.

Starting over from another standpoint, the designer then chooses a wing large enough to give him good takeoff and landing performance. Now when he calculates cruise speed, he finds it much lower than desired. Again he considers a larger engine, but finds the engine weight—plus the added fuel required by the larger engine—to add so much weight that his field length requirement has increased once more.

What usually happens in the end is the choice of a wing somewhere in between—one that will give reasonable speed and reasonable takeoff and landing performance. It is a compromise. Indeed, aircraft design is, by its very nature, a multitude of compromises. In the profession of design, this action is known as *tradeoff*. We trade performance in one area to gain performance in another. This activity is not confined to performance alone. Often we sacrifice performance to increase passenger comfort, reduce cost, improve maintainability, or increase safety. Mass-produced aircraft that will sell must be designed with all of these considerations in mind.

How often we hear the homebuilt designer berate the inefficiency of production airplanes. After all, his Super Duper II has an engine smaller than the lowest-powered production trainer, yet it will cruise faster than 200 miles per hour. It also lands at approximately 150 and requires the pilot to lie prone, as well as wear solid ear plugs to keep from going deaf. While these aspects might not bother him, they would deter the purchase of such airplanes by many others. To improve these qualities to the same degree as that of equivalent production aircraft would result in an airplane looking very much like all the others in that class, and performing about the same. There is usually a configuration that will give the best overall performance in a certain performance category. This is why airplanes of different manufacturers look so much alike, if they are to perform the same mission.

# INITIAL CONCEPTION

The first step in the conceptual phase of design is to study the specifications, or requirements. It was already pointed out that they will to some degree limit the configuration to certain types of powerplants, or to certain general physical size. A further study of the design specifications should be made to determine if there is any driving requirement in the design. This means an item that is critical or of high priority to the accomplishment of the primary mission. Top speed, for example, would be a driving requirement in fighter or racing airplane design. Range would be a driving requirement in a transoceanic transport. Such items would dictate the primary performance to consider in developing the concept; however, care must be taken not to get carried away by the driving requirements. You do not want to end up with a Mach 4 fighter that takes five miles to land, or a 10,000-mile transport that cruises at 100 knots.

Once you know the specifications and their relative importance in the design, you can then determine what characteristics to shoot for. Listed below are the major performance items and the characteristics that are desirable for best performance in the respective areas.

| Performance Item | Desirable Characteristics |
|---|---|
| Payload capacity: | Large size, high power, low weight |
| Speed: | Low drag (small component size), high power |

| Range: | Low weight, low drag, low fuel consumption, large fuel capacity |
|---|---|
| Takeoff distance: | High power, large wing area, high lift airfoil, low drag, low weight |
| Landing distance: | Large wing area, high lift airfoil, low weight |
| Climb rate and ceiling: | High power, low drag, low weight |

You can see some of the obvious conflicts in the characteristics listed. Long range, for example, is enhanced by low weight; however, a large amount of fuel is also required and this in itself adds weight. Large physical size is required to contain large payloads. Large size means higher structural weight, yet low weight contributes to more payload lifting capability.

Also, you will note that in many performance items there are requirements for the same characteristic to be low and another to be high. Low weight and high wing area are typical of such qualities. This leads to consideration of the ratio of these two characteristics. The ratio of weight to wing area (W/S, as it is usually written), is known as the *wing loading*, which represents the average weight that each unit of wing area must carry. In English units, this is measured in pounds per square foot (Newtons per square meter in metric). It is more meaningful in terms of performance to consider the ratios of these characteristics rather than the characteristics individually: thus, a 3,000-pound airplane with a 200 square-foot wing would have the same wing loading (W/S), as a 10,000 pound airplane with a 667 square-foot wing (namely, 15). The takeoff and landing performance of these two airplanes would be the same, with other things being equal. Such performance depends then not so much on the actual weight, but the weight to wing area ratio.

Another important ratio that you might spot in the above list is power to weight. High power is desirable in most cases, and also low weight; thus, like wing loading, the amount of power per unit of weight is a more meaningful quantity and is written as P/W, sometimes referred to as *power loading*.

Now it can be seen that load capability, speed, takeoff, and climb performance all depend on a high P/W ratio. Takeoff and landing require low W/S ratios, but for a given size airplane, high speed necessitates relatively high W/S ratios. We will see eventually how these ratios are used as we delve more into the procedures of airplane design.

The general field of airplane design is very complex and involves consideration of performance in many different flight regimes. In order to keep within the scope of this book, the remainder of the discussion will be limited to just one class of aircraft, light propeller-driven airplanes. The following considerations of various component configurations will thus apply to this type of aircraft. In developing the general concept, it is necessary to consider the advantages and disadvantages of different configurations of each major aircraft component.

## FUSELAGE DESIGN

If you have ever built model airplanes, you probably built the fuselage first. Most homebuilders, too, start with the fuselage. After all, that is the area that will become their personal space when completed. The wings, which carry the fuselage into the air, and the tail, which guides it, are usually secondary considerations and come later—at least in the mind of the builder.

The designer also seems to be intuitively motivated to begin with the fuselage. There is a more technical reason, though, for considering this component first. The primary mission of the airplane is to carry the payload specified and this payload, in most cases, must be contained in the fuselage. It must house one, two, four, or maybe 200 people, or so many pounds of cargo. This requirement will dictate the size of the fuselage and the general size of the overall airplane. To a large extent, it also determines the gross weight, at least in ballpark figures.

The natural inclination of the designer is, probably, to shape the fuselage for aerodynamic efficiency, meaning low drag. The shape that gives the lowest pressure drag at low subsonic speeds is shown in Fig. 8-1. This is an optimized aerodynamic shape, with circular cross section and with a *fineness ratio* of 3. Fineness ratio is the length (l) divided by the maximum diameter (d). Tests have shown that drag increases at fineness ratios greater than 3; drag increases even worse when the fineness ratio is fewer than 3.

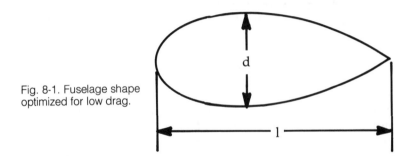

Fig. 8-1. Fuselage shape optimized for low drag.

In order to achieve this optimum shape, a typical four-place lightplane, with a fuselage length of approximately 24 feet, would have to have a diameter of 8 feet. A light twin with a typical length of 36 feet would require a diameter of 12 feet. Obviously this is much more room than required to seat human beings of current generations. If fuselages were actually shaped from this consideration alone there would be a lot of wasted space. More importantly, the additional surface area would generate unnecessary skin friction drag. The resulting overall parasite drag (pressure and skin friction combined) would not necessarily be the minimum possible.

A more practical fuselage design is a slight modification of the optimum shape. For transport aircraft, or light airplanes with a number of passengers, the cabin section is usually cylindrical (or near cylindrical) in shape. The cylinder is then rounded off in front and tapered to a point in the rear as shown in Fig. 8-2. This shape approaches that of the optimum configuration, but provides for a much more practical size and shape for the cabin area.

The circular cross section simplifies structural design and provides a convenient seating arrangement for transport aircraft as shown in Fig. 8-3. The area

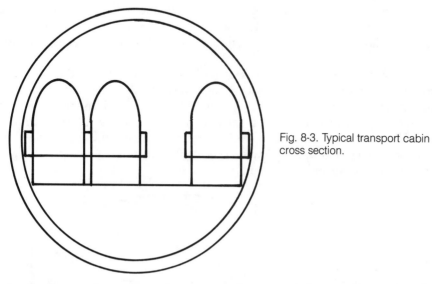

Fig. 8-2. Fuselage with cylindrical cabin section.

Fig. 8-3. Typical transport cabin
cross section.

under the floor makes a great place to store baggage and cargo. For two- or four-place lightplanes, where such space is not needed, the bottom of the circular section is usually cut off, forming the general cross-sectional shape as shown in Fig. 8-4. This shape is actually for a low-wing configuration. For a high-wing aircraft, a typical configuration would be this same general shape inverted.

Fig. 8-4. Typical lightplane cabin
cross section.

There are other factors to consider in the fuselage layout. In the conventional arrangement, the aft end of the fuselage supports the tail. The length must be sufficient to place the tail at a reasonable moment arm from the CG in order to give adequate stability. Otherwise, a very large tail would be required, creating excessive drag. In single-engine airplanes, the forward end of the fuselage (usually) houses the engine. There must be sufficient space for the engine and it must be placed far enough ahead of the CG to ensure adequate balance. Other than considerations such as these, the most practical approach to fuselage design is simply the smallest shell that will fit around the occupants. This approach provides for the minimum wetted area and thus the lowest skin friction drag. Of course, the shape cannot be too unwieldy or unusually high pressure drag will result.

Because the design is based on containing the occupants, seating arrangement becomes a prime consideration. In the case of two-place aircraft, the designer has the choice of *tandem* or *side-by-side* arrangement. Tandem seating has an advantage from an aerodynamic standpoint (lower frontal area), but the inconvenience in communication between the occupants has made this arrangement somewhat obsolete. Side-by-side seating is pretty much standard, even in two-place trainers. While the smallest space in which you can squeeze two people is desirable aerodynamically, in reality it is not too practical. Passengers and crew must have adequate space in which to feel comfortable. Certain dimensions in seating have been arrived at to ensure comfort for the average individual. Seat width varies from a minimum of approximately 16 inches to a maximum of approximately 22 inches. The difference depends on whether it is a short haul transport or trainer (in either case, the occupant will spend only approximately one hour onboard), or a long-range airplane (where passengers will sit for a number of hours). *Seat pitch*, or longitudinal distance between the same points on adjacent seats, also varies for the same reason from approximately 28 inches to 43 inches. Cabin height should be a minimum of approximately 4 feet for seated occupants, but approximately 5 feet or greater if passengers will move about the cabin in flight. In this case, some aisle space is also required.

# WING DESIGN

While the fuselage might be the part of the airplane of greatest concern to the passengers, the wing is certainly the most important to the aerodynamics of the craft. Aerodynamically, it is the heart of the airplane. Most of the aerodynamic behavior of the aircraft will depend on how the designer configures the wing.

## Basic wing configuration

In the conceptual phase, the designer has some very basic decisions to make about the wing. Probably the first of these is whether to make it a high-wing or low-wing design, or something in between. Aerodynamically, the high wing is somewhat superior to the low wing. The upper surface of an airfoil is the most critical and the uninterrupted contour of the high wing as it passes through the fuselage gives it a little better lift-to-drag ratio. The low wing has its critical upper surface disturbed in the fuselage area. There is more potential for interference drag to occur at the juncture of the low wing and the fuselage. This effect can be alleviated somewhat by proper *filleting* or *fairing*, as it is also known. Extensive filleting is employed on the Mooney wing-fuselage junction, as shown in Fig. 8-5.

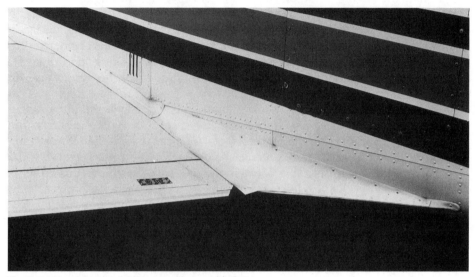

Fig. 8-5. Wing-fuselage fairing on a Mooney Mark 21.

The high wing also has inherent lateral stability, as discussed in chapter 6, and requires little or no dihedral. This results in slight improvement in L/D ratio of the wing. The low wing, on the other hand, encounters more ground effect (*see* chapter 3), and could result in decreased takeoff distance. Landing distance, however, is increased by ground effect, making the high wing more desirable from that standpoint. Midwing configurations, of course, have characteristics in between those of the high and low wings.

Wing choice is usually based on considerations other than aerodynamic, however. One major advantage of the low wing is to provide a structure to which the landing gear can easily be attached. With the low wing, the struts can be made relatively short and weight is thereby minimized. Weight, of course, ultimately affects the aerodynamics of the airplane, so in a roundabout way, this really is an aerodynamic consideration.

Because the wing normally houses the fuel, refueling and checking fuel supply is facilitated by a low-wing design; however, that fuel location presents more of a potential fire hazard in the event of a crash. Fuel tanks in the low wing are more likely to rupture than those in a high wing in a crash of limited severity. On the other hand, the low-wing structure serves as an energy-absorbing device, and could provide greater crash safety—provided that fire does not ensue.

Low-wing airplanes are also somewhat more maneuverable, and are preferred by aerobatic pilots (or those who fancy themselves to be aerobatic pilots).

While on the subject of aerobatics, we probably should make some mention of biplane configuration as a possible consideration. Biplanes have high drag characteristics, and are pretty much passe in the broad spectrum of aircraft use. They do have several advantages, however. One is the potential for a high roll rate due to the relatively short span. The other is a rather lightweight wing structure that can be built to be very strong. All of these characteristics are desirable in aerobatic craft. Biplanes could prove to be efficient for such use, where high speed is not a concern.

To sum up the discussion of high versus low-wing design, the following list contains the relative advantages of each:

*High Wing*
- Better L/D ratio
- Better lateral stability
- Shorter landing distance
- Better crash fire protection

*Low Wing*
- Better landing gear support
- Better roll maneuverability
- Easier refueling
- Shorter takeoff distance
- Better crash energy absorption

There are a few other, less technical, differences also. High wings are often preferred by certain pilots and passengers because they offer a better view of the ground. This might even be required in the case of spotting operations or aerial photography; however, low wings usually present a more aesthetically pleasing airplane design. Even without any aerodynamic advantage, many aeronautical aficionados will opt for the low wing simply because it is "neater."

Another choice that a designer must face has to do with structural considerations. Wings can be externally braced with struts or supported entirely by internal structure. The latter configuration is referred to as a *cantilever* structure. External bracing was relied upon in olden days, when the art of structural design was not too well developed. When high strength-to-weight aluminum alloys came along, most designers gravitated toward the cantilever wing configuration. Such arrangement eliminates the drag of the external struts; however, the amount of structure required for internal support results in a heavier wing. The strut-braced wing of the same size is lighter, and often the weight savings outweighs the slight increase in drag. This is why such airplanes as the venerable Cessna 172 (Fig. 8-6) have persisted with the somewhat old-fashioned-appearing struts tucked under its wings.

Struts are normally only used on high-wing designs. There are two good reasons for this: First of all, the strut below the wing is in tension under normal flight loads on the wing. Tensile loads require less structure (even wires can carry tension). A strut on a low wing would be in compression and would require a

Fig. 8-6. Cessna 172 Skyhawk.

hefty amount of material to keep it from buckling. Secondly, the strut-wing juncture offers more interference, and hence, more drag with a strut on the upper surface than it does on the lower surface. Making the cantilever or strut-braced wing decision is thus tied in with the choice of high or low wing.

## Planform selection

Although the exact shape of the wing is not usually considered in the basic conceptual phase, we will discuss this aspect while on the subject of wings. *Planform*, as first mentioned in chapter 2, is the shape of the wing as viewed from directly above (or below, depending on your mental point of view). Chapter 3 explained why the elliptically shaped wing is considered the ideal wing. It has the lowest induced drag of any shape wing producing the same amount of lift. Elliptical wings are difficult to produce (translated "expensive" to the consumer) and are usually overlooked in place of tapered wings, which have almost the same aerodynamic characteristics.

An even more economical wing is the rectangular shaped wing. This wing is the cheapest to produce because its profile shape is exactly the same throughout the span and all of the ribs are thus identical and can be made from the same pattern. The problem with the rectangular wing is that it ends up being much heavier than necessary in the outboard portions. This is because the lift load decreases toward the tips, as shown in Fig. 2-38, and hence not as much structure is required.

Stall progression, however, is much more favorable for the rectangular wing, as explained in chapter 2 and illustrated in Fig. 2-41. Tapered wings can be forced into good stall patterns by twist (washout) of the outboard sections; however, this results in increased parasite drag. Tapered wings, on the other hand, increase the aspect ratio over a rectangular wing for the same given wing area, and thus reduce the induced drag. A good compromise on planform is a combination of both rectangular and tapered configurations, as shown in Fig. 8-7. The inboard portion is rectangular, allowing for both low production cost and good stall characteristics. The outboard portion is then tapered from a certain spanwise station to reduce weight and increase aspect ratio. This arrangement was adopted by Cessna and Piper in single-engine designs.

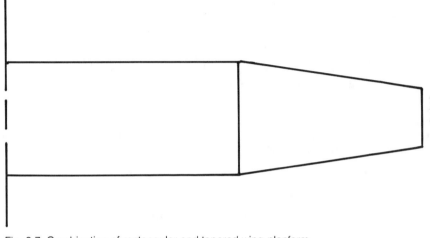

Fig. 8-7. Combination of rectangular and tapered wing planform.

## Airfoil selection

The profile shape of the wing is determined by the choice of airfoil. Choosing an airfoil is a more complex decision than arranging the planform. The many, many different airfoil shapes available can present a confusing situation to the neophyte designer. A closer look at airfoil types reveals that there are relatively few families of airfoils, such as NACA four-digit or NACA 6-series airfoils. Within each family, the airfoils vary according to thickness, degree of camber, or a few other properties. Remember that these characteristics are always given as a ratio of the chord length, so that the shape can be scaled to any desired size while retaining the same properties.

Primarily, we normally seek an airfoil that gives high lift and low drag, just as we seek to "buy low and sell high." As in economic situations, however, we cannot always achieve this ideal situation. Compromise is again the key word. What we are really attempting to attain in this respect is the highest ratio of lift to drag. We also want to have this high L/D to occur at some critical condition, usually cruise speed. Such a critical condition is referred to as the *design point*. We attempt to optimize the design for this condition and must put up with somewhat less than optimum efficiency at other conditions, or "off design" points.

Another major consideration is the maximum attainable lift coefficient. High maximum $C_1$ means low stalling speed. Also, the sharpness with which the lift coefficient curve drops off at its maximum point (stall point) will determine how abruptly the wing will stall. Thin low-camber sections are the worst in this category. Also important is the pitching moment, as discussed in chapter 2. Higher pitching moment means a higher balancing tail load, which leads to increased drag from the horizontal tail. Other nonaerodynamic, but important, considerations include such things as sufficient thickness to provide for the spar structure, fuel containment, and landing gear housing (if retractable) and a shape that facilitates production.

The optimum airfoil should therefore possess the following properties:

- Low drag coefficient ($C_{do}$)
- Minimum drag at design lift coefficient
- Maximum lift coefficient ($C_{1_{max}}$) as high as possible
- Pitching moment coefficient ($C_{mac}$) as near zero as possible
- Sufficient thickness for spar, fuel, and landing gear

The two properties that have the most effect on these characteristics are thickness and camber. The effects of increases in each of these two properties can be summarized:

*Increased Thickness*
- Increases maximum lift coefficient (up to approximately 18 percent)
- Increases drag coefficient
- Provides greater space for structure and fuel

*Increased Camber*
- Increases design lift coefficient
- Increases (in negative value) pitching moment
- Increases lift coefficient at a given angle of attack

The appendix includes published data from tests on several NACA airfoils. On the left side of the charts is a plot of lift coefficient versus angle of attack. The higher of these curves is for a split flap extended $60°$. On the right-hand chart is a plot of section drag coefficient versus lift coefficient, and also pitching moment versus lift coefficient in the lower portion of this chart. Different symbols are used for different Reynolds numbers, or different size and speed regimes. An airplane going 150 miles per hour with a 5.5-ft. chord wing operates at a Reynolds number of approximately 6,000,000 (written as $6 \times 10^6$ in the charts).

Note the differences between the 1412 airfoil and the 4412. Both have the same thickness (12 percent), but the 4 percent camber of the 4412 gives it a much more negative pitching moment. The minimum drag also occurs at a $C_1$ of 0.4, while the 1412 has its minimum drag at approximately 0.1. Notice also the much more gradual drop of the lift curve for the 4412, while the 1412 stalls very sharply. Both airfoils, having the same thickness, have approximately the same minimum drag. Notice the increased drag of the 4415, an airfoil nearly identical to the 4412, except slightly thicker.

Again the designer has a dilemma: Thickness, for example, is needed for structure and fuel, but hurts drag. Experience has shown that airfoils of approximately 12 to 15 percent thickness ratio work best for most light airplanes. Much less than 12 results in either a very heavy wing, or one that cannot be internally supported. More than 15 percent results in excessive drag. Also, 2 percent camber seems to be approximately average, with 4 percent used in some cases; however, somewhere in this range is a good compromise on camber for best overall results.

Thickness to give adequate spar depth cannot be overemphasized. Most of the wing weight comes from the bending member support, which is usually a spar. The spar is given most of its strength by its depth. The weight of the spar is actually inversely proportional to the square of its depth. Limiting the depth only slightly means that it must be much wider and consequently heavier.

Another consideration in airfoil choice is the location of the maximum thickness point. This is usually where the spar is placed to take advantage of maximum depth capability. The spar has to pass through the fuselage from one wing to the other. On a low-wing airplane, it is highly desirable to have this occur under a seat. In many four-place airplanes, an airfoil with a maximum thickness point far aft is chosen largely to place the main spar under the rear seat.

Spar location might seem like a trivial matter, but it can be extremely aggravating to alter the design later if the spar ends up where the pilot's feet ought to be. The Boeing 247, the first all-metal, low-wing, retractable gear transport, built in the early 1930s, had the main spar run directly through the passenger aisle. Passengers had to scale and descend several steps to pass over it going from one end of the cabin to the other. This points out another consideration in the choice of high or low wing, and, for many passenger arrangements, pretty much rules out the midwing design.

# POWERPLANT SELECTION

Just as the wing is the primary (and often sole) provider of lift, the powerplant gives the airplane its thrust; hence, these two components provide the positive forces necessary for flight. Obviously, then, the powerplant selection is another very important consideration of the designer.

Engine characteristics and the relative advantages of various types are discussed at length in chapter 4. There appears to be a definite basic type of engine that is best for each speed regime of flight. For the slow-speed light airplane, the designer can very quickly zero in on a piston engine-propeller powerplant. All modern piston engines are of the air-cooled, horizontally-opposed cylinder design. The choice is thus pretty much limited to a determination of the specific engine model that develops the power required.

Power-to-weight ratio is an important characteristic. If there are several engines of the same power rating, the highest P/W ratio might determine the selection; however, most modern engines of the same basic power class have approximately the same power-to-weight ratios, on the order of 0.5 horsepower per pound. Another very important consideration is specific fuel consumption or how many pounds of fuel are required per hour to develop one horsepower. Again, most engines have been refined to the point that nearly all of the same basic type have approximately the same specific fuel consumption. For modern reciprocating engines this figure is approximately 0.42 to 0.50 pounds per hour per brake horsepower.

The actual choice between different models with the same power rating usually amounts to consideration of lesser details. Such things as number and location of accessory drives, provision for controllable pitch prop, or availability with fuel injection, gearing, or turbocharging, might be deciding factors. Geared engines can increase power by running at very high rpm. Fuel injection increases power slightly, but more importantly, improves fuel efficiency at low engine speeds. Turbocharging increases altitude capability, which usually results in considerable improvement in airplane performance; however, all of these features increase the engine weight to some extent, and can increase cost to a considerable extent.

The actual engine dimensions and the location of the carburetor are important for the cowling design and will have a considerable influence on the shape and size of the forward end of the fuselage. The thrust line must be located so that it can be situated near to the vertical location of the CG to minimize pitching moment changes with power changes.

Propeller characteristics are also covered in chapter 4. The manufacturers can recommend the best choice of propeller for a given engine and airplane operating conditions. Basically, this involves selecting the pitch and diameter to yield maximum efficiency at the design condition and reasonable efficiency over the rest of the flight envelope. The designer does have the choice of controllable or fixed-pitch. Controllable-pitch props extend the range of maximum efficiency and can improve climb performance while retaining high cruise speed. They do add weight and cost to the airplane, however. On very small airplanes these factors might outweigh the advantages of improved performance.

Certain regulations must also be observed in propeller installation. For example, current FAA regulations require a ground clearance of at least 7 inches for tricycle-geared airplanes, or 9 inches with tailwheel types. This applies to static deflection of the landing gear, taxiing, and level takeoff situations.

## LANDING GEAR CONFIGURATION

For many years the tailwheel-type landing gear was the only landing gear. For this reason it is often regarded as old fashioned or obsolete, and shunned by modern pilots and designers alike. The tailwheel was probably popular for so long largely because it was simple to make and install. It also added very little weight and drag to the airplane. Also, with tailwheel configurations, the main gear, being ahead of the CG, requires a location that usually falls right under the heaviest part of the wing. This makes the attachment of the gear to the wing or fuselage relatively easy.

Tailwheel configurations, however, complicate landing and taxiing operations. With the CG behind the main gear, a premature landing impact causes the airplane to rotate to a higher angle of attack and the resulting lift increase causes it to bounce back into the air. Also, any yawing tendency (from braking or crosswind) is amplified by the aft-located CG and embarrassing (or even damaging) ground loops can be the result of only a slight bit of inattention from the pilot. In addition, the pilot's vision of the runway or taxiway ahead is almost totally obliterated by the nose-high position of the taildragger at very slow speeds.

All of these disadvantages of the tailwheel configuration can be eliminated with the nosewheel, or tricycle gear, configuration. In this case the CG is ahead of the main gear and it exerts a stabilizing influence on pitch and yaw motion. A hard impact on the main gear causes a pitch-down motion, decreasing lift. Likewise, a yawing tendency is counteracted by the forward CG.

Unfortunately, the nose gear has to withstand considerable impact itself and thus ends up being pretty heavy. Its size and location on the critical forward end of the airplane add a fair amount of drag. Tricycle gear configurations also place the main gear fairly far aft and complicate attachment to the rest of the airplane. Nevertheless, the greatly improved handling qualities and the greater visibility (becoming more and more essential as traffic increases) have caused a mass adoption of this type of landing gear. Into the 1980s, only a few special purpose airplanes, such as agricultural craft or utility aircraft like the Cessna 185 (Fig. 8-8) were manufactured with tailwheels. These aircraft often operate from unimproved landing strips where nosewheels would not hold up.

Cessna Aircraft Company

Fig. 8-8. Cessna 185 Skywagon.

# TAIL DESIGN

The purpose of the tail surfaces is to provide adequate stability and control. The size of these surfaces is thus determined by the degree of stability and/or control desired. The horizontal tail provides longitudinal stability and control and the vertical tail gives these qualities in the directional sense. The total horizontal tail surface provides longitudinal stability, while the elevator provides pitch control. The same is true of the vertical tail and rudder for yaw stability and control, respectively.

Aerodynamic forces are proportional to the area of a surface; thus, a larger tail surface will produce more lifting force on the tail. Stability is determined by the moments about the CG and a moment requires a moment arm; hence, the distance, or moment arm, from the CG to the tail surface is also a factor. The tail moment that provides longitudinal stability is actually the tail lift force times the tail moment arm as shown in Fig. 6-1. Because the force is determined to some extent by the area, stability will be proportional to tail area times the moment arm.

Area has units of length squared (such as square feet) and, when multiplied by another length, has units of length cubed (such as cubic feet). Cubic length is also the dimensions of a volume, so that the product of tail area times the tail arm is referred to as *tail volume*. Stability is proportional to this volume, so that an increase in either tail area or tail arm will increase stability; thus, a long fuselage would create a long moment arm and would require less tail area. A short fuselage, with corresponding short tail arm, would necessitate a larger tail for the same degree of stability.

Tail volume *coefficients* are defined, analogous to lift or drag coefficients, by dividing the tail volume by another appropriate volume. In the case of the horizontal tail, the wing area and average chord dimensions largely determine the amount of stability required from the tail; hence, the *horizontal* tail volume coefficient is defined as the actual tail volume divided by the product of wing area times average wing chord.

$$\text{Horizontal tail volume coefficient} = \frac{\text{Horizontal tail area} \times \text{tail arm}}{\text{wing area} \times \text{wing chord}}$$

In more technical use, this is usually abbreviated:

$$\overline{V}_H = \frac{S_h \times l_t}{S_w \times c}$$

The symbols stand for the respective terms as spelled out in the equation above. A similar coefficient can be defined for the vertical tail, except that here the wing area and the wing span are the important quantities; thus, the *vertical* tail volume coefficient is written:

$$\overline{V}_v = \frac{S_v \times l_t}{S_w \times b}$$

Light airplanes have horizontal tail volume coefficients ranging from approximately 0.3 to approximately 0.7, depending upon the amount of stability or control required. Large degrees of flaps, which yield high lift coefficients, require greater horizontal tail surface for control purposes and would require the higher tail volumes. As an example of tail volume coefficient, consider a Piper Arrow II. This airplane has the following dimensions:

| | |
|---|---|
| Horizontal tail area: | 32 sq. ft. |
| Tail moment arm: | 14 ft. |
| Wing area: | 170 sq. ft. |
| Average chord: | 5.25 ft. |

The tail volume coefficient is:

$$\overline{V}_H = \frac{S_h \times l_t}{S_w \times c} = \frac{32 \times 14}{170 \times 5.25} = 0.5$$

Tail areas are estimated in the early stages of design by determining the tail volume from a *typical* tail volume coefficient. For example, suppose we wanted a degree of stability approximately equal to the Arrow mentioned above; however, our new airplane design has a wing area of 240 square feet and a chord of 6 feet. Using the Arrow's tail volume coefficient, we multiply by our design's wing dimensions so that the required tail volume is found as follows:

$$\text{Tail volume} = \overline{V}_h \times S_w \times c = 0.5 \times 240 \times 6 = 720 \text{ cu. ft.}$$

We could use a tail area of 45 sq ft. if the tail arm could be 16 feet. If this length arm is not available (due to planned fuselage length) and had to be limited to 15 feet, the tail area would have to be 48 square feet. Either combination would yield 720 cubic feet total tail volume.

Vertical tail sizes are estimated in a similar way. Vertical tail volume coefficients range from approximately 0.02 to approximately 0.05 for typical light airplanes. Such estimates are sufficient for preliminary estimation; however, detailed design of the tail for overall stability and control is quite complex and beyond the scope of this book.

## Tail configuration

The conventional tail arrangement is the single vertical fin mounted above the horizontal stabilizer as shown in Fig. 8-9. If the vertical fin requirement results in a very large fin, significant rolling moments can result from rudder deflection. This can be overcome by use of twin fins. Each of these is smaller, but the total area of both contribute to directional stability. There is some additional interference drag at their junction with the horizontal surface, but the endplate effect they have on the horizontal stabilizer makes it more efficient. It could thus be made smaller and reduce some drag in that way. Twin fins as used on the Ercoupe are shown in Fig. 8-10.

Fig. 8-9. Conventional vertical tail on a Cessna 310.

Fig. 8-10. Twin vertical fins on an Ercoupe.

Nothing is gained from the swept vertical tail on a low-speed airplane except a snazzy look. The sweep detracts from efficiency at slow speeds just as it does with a wing. This in turn requires more vertical tail area and, consequently, creates a bit more drag. Sales of such configurations proved that a swept tail is worth the tradeoff because airplane buyers seem to relish aesthetics as much as efficiency.

Another consideration for tail arrangement has to do with spin recovery. Tests have shown that horizontal tails mounted below the vertical tail tend to blank out the vertical tail from the airflow in a spin, making recovery difficult or impossible. Figure 8-11 shows this effect due to the direction of the airflow in a spin. Note that the swept tail is a particular offender in this aspect of flight. Improvement in spin performance can be obtained by moving the horizontal stabilizer farther aft, placing it up on the vertical fin, or extending the fin below the horizontal stabilizer. All of these arrangements complicate mounting somewhat and usually result in a requirement for heavier structure.

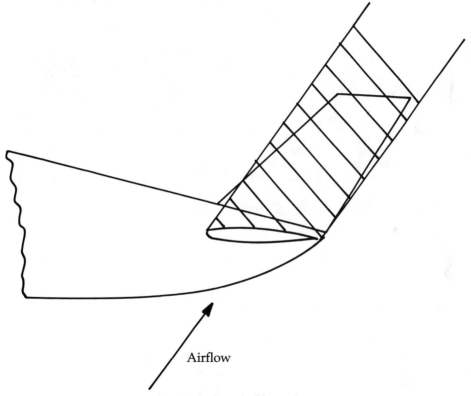

Airflow

Fig. 8-11. Blanking of vertical tail by the horizontal tail in a spin.

## T and V-tail

One of the best configurations for spin recovery is the T-tail, shown in Fig. 8-12 and also in Fig. 2-33. T-tails gained considerable popularity in the early 1980s. This was due partly to considerable research on spin recovery conducted by NASA in the late 1970s that pointed to this configuration as ideal for spin recovery. Another reason for using the T-tail is to place it out of the wing's downwash. Downwash reduces the stabilizing effect of the horizontal tail and this effect becomes greater at high angles of attack. The low-mounted tail is immersed in this downwash as shown in Fig. 8-13A, while the T-tail remains free of it and

Fig. 8-12. T-tail arrangement of the Grob G109 powered sailplane.

retains its effectiveness. However, at full stall the downwash ceases and the wing's wake flows pretty much directly aft. In this situation the low-mounted tail is free of the wake, while the T-tail often falls right in it, as shown in Fig. 8-13B. If this happens, the T-tail experiences a sudden loss of effectiveness, and a rapid pitch-down motion results, sometimes referred to as a *deep stall*.

Another disadvantage of the T-tail is the additional weight required for the heavier structure necessary to support the horizontal tail in this position. Elevator control and trim linkage are also complicated and add additional weight. The T-tail does improve the effectiveness of the vertical tail by its endplate effect, similar to the effect of twin vertical tails on the horizontal tail. Much of the shift to T-tails has been due to the modern appearance of such configurations. The enthusiasm for the T-tail, like the swept tail, is usually greater in the sales department than in the engineering department. Some manufacturers are finding that the disadvantages of this design outweigh the advantages. Witness Piper, who discontinued the T-tailed Lance, and put the tail on the Saratoga, the last version of the six-place Cherokee design, back on the bottom, where it was originally.

Another tail configuration that keeps popping up is the V-tail. In this design, a single surface on either side of the centerline is canted upward to provide horizontal and vertical tail effects. The vertical projection of the V-tail provides longitudinal stability and the horizontal projection provides directional stability. This

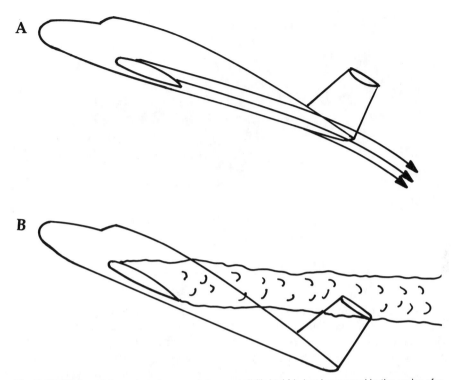

Fig. 8-13. T-tail avoiding wing downwash in prestall flight (A), but immersed in the wake of a stalled wing (B).

arrangement reduces the drag slightly over that of a conventional tail arrangement. The most famous V-tailed aircraft is undoubtedly the highly successful Beechcraft Bonanza (Fig. 8-14), which was somewhat of a revolutionary light airplane when first introduced in 1947. Many thought that its high performance was due largely to the V-tail; however, this aspect contributed only slightly to the overall low-drag configuration. This is apparent if you look at the straight-tailed version of the Bonanza (Model 33), which has almost identical performance to that of the V-tail model when the same engine is installed.

The drag reduction in the V-tail arrangement comes about only from the savings in interference drag due to fewer junctures of the surfaces. The total surface area of the V-tail must be equal to the total of a conventional vertical and horizontal tail in order to provide the same degree of stability. There is therefore no savings in skin friction drag.

The main objection to the V-tail is the extremely complicated control system required to get pitch and yaw control from a single control surface. The V-tail is also susceptible to considerable Dutch roll tendencies, as discussed in chapter 6. Both V-tails and T-tails have become popular on sailplanes, but this is primarily to keep the tail surfaces high so that they are not damaged in landing on unimproved terrain. The V-tail also shares honors with the T-tail in having good spin recovery characteristics.

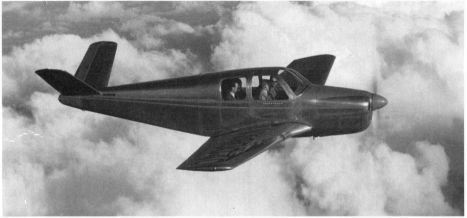

Fig. 8-14. Original 1947 Model Beech Bonanza.

# FIRST ESTIMATIONS

At first it seems almost impossible to begin the design of an airplane. Everything that you need to know is unknown. For example, in order to determine the weight of the airplane, you need to know the size of the engine and amount of fuel required. Before you can determine the engine size, however, you must know how much drag has to be overcome. The drag, in turn, depends on the weight. So you are back to square one. There seems to be no logical starting point.

To resolve this dilemma, we resort to a strategy of estimating certain parameters initially, to form a starting point for the design. At first this might seem like making a wild guess and then seeing how close you can come to it; however, there is a fairly legitimate and generally accepted method for accomplishing this feat. The trick is to examine the values of concern in airplanes of existing design. They must, of course, be in the same general class of size and performance. Certain characteristics are revealed to be quite consistent from one airplane to another, and these are obviously the characteristics to start with. Exact values for the new design are not chosen at this stage, but only average values or "ballpark" figures, as they are often referred to.

## Weight

Probably the best place to start is to make a first estimate of the gross weight. So many other parameters depend on this value that very little else can be done in the design process without it. Figure 8-15 shows the various steps in a logical design sequence, with weight estimation comprising the first step. The sequence as shown is not the only way to proceed, but it is one followed by many designers, and the one that we will use in our discussion.

The takeoff gross weight is made up of the empty weight of the airplane (including engine and airframe), the weight of the fuel (total fuel at takeoff), and the payload. In equation form:

$$W_G = W_e + W_f + W_p$$

where  $W_G$ = takeoff gross weight
       $W_e$ = empty weight
       $W_f$ = fuel weight
       $W_p$ = payload weight

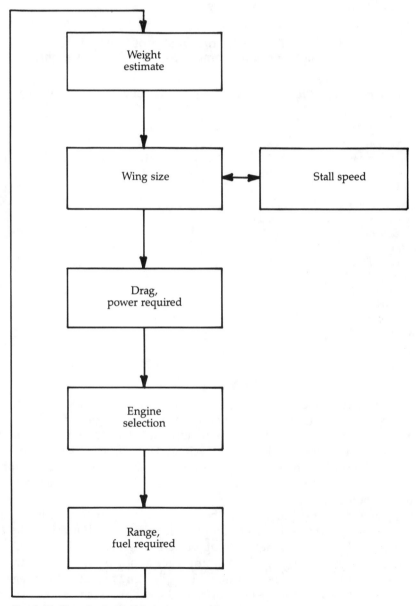

Fig. 8-15. Flowchart of initial design procedure.

For light airplanes certified in Normal category and transport airplanes, the empty weight is a fairly consistent proportion of the gross weight. For airplanes of conventional metal semimonocoque construction, this figure is very close to 55 percent from small 4-place singles up to large jets. A closer look at light single-engine airplanes shows that this figure actually ranges from approximately 0.54 to 0.62 times the gross weight. Table 8-1 shows 16 single engine airplanes with gross weights from 2,400 pounds to 4,000 pounds and their respective portion of gross weight making up the empty weight.

**Table 8-1. Portion of gross weight of light aircraft in empty weight and fuel weight.**

| Aircraft | Gross Wt. (lbs) | Empty Wt. ratio $\left(\dfrac{W_E}{W}\right)$ | Fuel Wt. ratio $\left(\dfrac{W_F}{W}\right)$ |
|---|---|---|---|
| Fixed Gear: | | | |
| Cessna 172 | 2,400 | 0.60 | 0.16 |
| Piper Warrior | 2,440 | 0.55 | 0.12 |
| Beech Sundowner | 2,450 | 0.61 | 0.14 |
| Piper Archer | 2,550 | 0.55 | 0.11 |
| Piper Dakota | 3,000 | 0.54 | 0.14 |
| Cessna Skylane | 3,100 | 0.56 | 0.17 |
| Cessna 206 | 3,600 | 0.54 | 0.15 |
| Piper Saratoga | 3,800 | 0.54 | 0.17 |
| Retractables: | | | |
| Cessna Cutlass | 2,650 | 0.60 | 0.14 |
| Mooney 201 | 2,740 | 0.61 | 0.14 |
| Beech Sierra | 2,750 | 0.62 | 0.12 |
| Piper Arrow | 2,900 | 0.58 | 0.15 |
| Cessna Skylane RG | 3,100 | 0.57 | 0.17 |
| Beech Bonanza | 3,400 | 0.62 | 0.13 |
| Cessna 210 | 3,800 | 0.57 | 0.14 |
| Cessna 210T | 4,000 | 0.56 | 0.13 |

Also shown in Table 8-1 is the amount of the gross weight that is taken up by the fuel. Notice that this figure is also fairly consistent, ranging from 0.11 to 0.17 for the aircraft listed. An average value of approximately 15 percent would be a good estimation for this portion of the weight. This value goes up with longer range aircraft or turbine powered airplanes. Small turboprop twins carry approximately 20 percent of the gross weight in fuel, and large, long-range jets have as much as 40 percent.

The payload is something that can be determined rather accurately. FAA regulations require everyone to weigh 170 pounds (not really, but this is the standard weight established for determining design weight, as well as weight and balance calculations during operations); thus, for a four-place airplane, the payload (pilot included) would be four times 170, plus any amount of baggage chosen to be included. With this value known, and good guesses possible for empty weight and fuel weight, the gross weight can be estimated. This process is best shown by equation:

$$W_G = \frac{W_P}{1 - \dfrac{W_e}{W_G} - \dfrac{W_f}{W_G}}$$

Using 0.55 as a value for $W_e/W_G$ and 0.15 for $W_f/W_G$, the equation becomes:

$$W_G = \frac{W_P}{1 - 0.55 - 0.15} = \frac{W_P}{0.3}$$

In other words, the payload divided by 0.3 would yield the gross weight to consider as a first estimate.

Note that this figure is for Normal category airplanes only and for those of conventional construction. Aircraft certificated in Utility category must be stronger and consequently have heavier airframes. A figure closer to 0.66 times the gross weight is more accurate for empty weight estimation in that case. For acrobatic category, the empty weight again goes up to approximately 72 percent of gross weight.

## Wing

With a weight estimation determined, the designer can proceed to choose the wing dimensions. The wing loading or weight to wing area ratio (W/S) is pretty much proportional to the cube root of the gross weight (written $\sqrt[3]{W}$). The average wing loading actually follows the equation:

$$\frac{W}{S} = 2.24 \times (\sqrt[3]{W} - 6)$$

for W/S calculated in pounds per square foot. For example, consider an 8,000 pound airplane. The cube root of 8,000 is 20.

$$\frac{W}{S} = 2.24 \times (20 - 6) = 31.4 \text{ lbs/sq. ft.}$$

It has also been found that wing loading will vary approximately 30 percent either way from this average figure, depending on whether the airplane is designed primarily for high-speed or low-speed operation; thus, W/S values from approximately 22 to approximately 41 lbs/ft. would not be unreasonable for an 8,000-pound airplane. Dividing the gross weight by the wing loading would then determine the wing area (S) required.

$$S = \frac{W}{(W/S)}$$

At this point another aspect of performance must be considered. Higher wing loading leads to higher stall speeds and correspondingly longer takeoff and landing distances. Wing loading and area from this standpoint must also be calculated to see what can be tolerated. In terms of stall speed, W/S is defined as follows:

$$\frac{W}{S} = C_{L_{max}} \times \frac{1}{2} \times \varrho \times V_s^2$$

At sea level, for stall speed ($V_s$) in knots, this equation becomes:

$$\frac{W}{S} = .003396 \times (C_{L_{max}}) \times (V_s)^2$$

FAR Part 23 requires single-engine airplanes to stall at 61 knots or less. With this requirement as a limiting factor, the wing loading equation is:

$$\frac{W}{S} = 12.23 \times (C_{L_{max}})$$

With no flaps, a maximum lift coefficient of 1.3 is approximately as much as can be expected. Slotted flaps can increase this value to almost 2.0, and exotic flap configurations are required to go much higher; therefore, for single-engine airplanes with slotted flaps (the normal type used in modern airplanes), a value of 25 pounds per square foot is approximately the maximum tolerable wing loading. For most modern lightplanes the wing loading is fewer than 20 lbs/sq. ft.

The wing area can thus be determined from the above considerations of wing loading. Next the designer must select an aspect ratio (AR). These values range from approximately 6 to 8 on light airplanes, although some twins have aspect ratios as high as 12. The average value of current production singles seems to be approximately 7.3. Once this choice is made, the span (b) can be determined. It is the square root of the wing area multiplied by the aspect ratio:

$$b = \sqrt{S \times (AR)}$$

## Power

Now the designer has nearly everything needed to make a drag estimate. Chapter 3 has an equation for the total drag of an airplane, breaking it into parasite and induced drag terms. Chapter 5 explained that drag is often measured in terms of power required and that power required is simply drag times velocity. The actual equation for power required in terms of horsepower, called thp for thrust horsepower, is:

$$thp_{req} = \frac{1}{550} \times \left[ C_{D_P} \times \frac{1}{2} \times \varrho \times S \times V^3 + \frac{2}{\pi \times \varrho \times e} \times \left( \frac{W}{b} \right)^2 \times \frac{1}{V} \right]$$

At this point we have estimated W, S, and b. The density, $\varrho$, depends on the altitude selected, and the value of V depends on what speed we want to consider. Usually, power is determined by what is needed for cruise speed, so the values of $\varrho$ and V could be selected for cruise altitude and velocity, respectively. The only values in the equation that we do not yet know are the parasite drag coefficient, $C_{Dp}$ and the Oswald efficiency factor, e. Again, experience has shown that for light retractables, $C_{Dp}$ ranges from approximately 0.020 to 0.030 and for fixed-gear airplanes from 0.025 to 0.040. Good first estimates are 0.025 for retractables and 0.035 for fixed-gear types. The value of e depends on wing location and is approximately 0.6 for low wings and 0.8 for high wings.

With all of these values known or estimated, they can be plugged into the above equation and the thrust horsepower required ($\text{thp}_{req}$) can be calculated. Now we select an engine—or at least the required power output of the engine. Remember that thrust horsepower is brake horsepower multiplied by propeller efficiency; thus, brake horsepower required ($\text{bhp}_{req}$) is thrust horsepower divided by propeller efficiency. Usually, a propeller is capable of 80 percent efficiency at cruise speed, so 0.8 is a good value to use for this figure; hence:

$$\text{bhp}_{req} = \frac{\text{thp}_{req}}{\eta_p} = \frac{\text{thp}_{req}}{0.8}$$

However, this figure of bhp is for cruise condition. Usually a cruise setting is a maximum of 75 percent power; therefore, the rated power must be even greater than this, or in this case, required bhp divided by 0.75.

$$\text{bhp}_{rated} = \frac{(\text{bhp})_{req \text{ for cruise}}}{0.75}$$

## Range

With the rated horsepower thus determined, you could, at this point, tentatively select a specific engine. An accurate fuel consumption figure would then be available from the engine specifications; however, even without a specific engine selection fuel consumption can be estimated. Modern piston engines consume from approximately 0.42 to 0.50 pounds of fuel each hour for each horsepower developed. This is known as the specific fuel consumption, SFC. Often a value of 0.5 is used for estimating purposes, to be on the conservative side. The total fuel available is determined by multiplying the fraction times the gross weight (approximately 0.15 for light airplanes).

$$\text{Total fuel} = \frac{W_f}{W_G} \times W_G$$

The range can now be determined by multiplying this figure by the cruise speed and dividing by the SFC times the bhp:

$$\text{Range} = \frac{(\text{Tot. Fuel}) \times V_{cr.}}{(\text{SFC}) \times (\text{bhp}_{req})}$$

This, of course, is absolute range and does not allow for any reserve or additional fuel required for takeoff and climb. It also does not consider the reduction in weight as fuel is burned off, which actually increases the range somewhat beyond this figure. However, it is reasonable for a first estimate. If the range is not adequate to meet the original specifications, then the amount of fuel required to attain the desired range can be computed. Because this will now increase the gross weight, a new weight estimate must be made, and the whole procedure is cycled through again. This process is referred to as *iteration*, and is the basis for refining the design parameters to achieve the best compromise in all-around performance.

## Initial estimation example

In order to better understand the initial design estimation procedure, let us consider an example. Let us assume that an airplane is to be designed to the following specifications:

- Payload: Four persons (including pilot) plus 20 pounds baggage each.
- Cruise speed: 150 knots.
- Range: 800 nautical miles.
- Certifiable under FAR Part 23.

In order to achieve the cruise speed, the designer opts for a retractable gear. To facilitate gear installation and retraction, he decides on a low wing: however, to keep production costs down he decides on a rectangular wing, which requires a lower aspect ratio for weight control. He decides that he can achieve a structural weight that will result in a $W_e$ of $0.55\ W_G$ and that the drag can be kept to a $C_{Dp}$ of 0.025. He also decides to shoot for 7,000 feet as a cruise altitude because most engines deliver 75 percent power up to approximately this level. Thus, he makes the following choices or assumptions:

$W_e = 0.55\ W_G$
$W_f = 0.15\ W_G$
$C_{DP} = 0.025$
$AR = 6$
$\varrho = 0.001927\ (7,000\ \text{ft})$

Proceeding then through the various steps of Fig. 8-15, the following calculations and determinations are made.

1. *Weight Estimation*
   Payload weight: $W_p = 4 \times (170 + 20) = 760$ lbs.
   Gross weight:

   $$W_G = \frac{W_p}{1 - \dfrac{W_e}{W_G} - \dfrac{W_f}{W_G}} = \frac{760}{1 - 0.55 - 0.15} = 2533 \text{ lbs}$$

2. *Wing Size*

   Area: $\sqrt[3]{W} = 13.6$

   $$W/S = 2.24 \times (\sqrt[3]{W} - 6) = 2.24 \times (13.6 - 6) = 17$$

   $$S = \frac{W}{W/S} = \frac{2533}{17} = 149 \text{ ft}^2$$

   Span: $b = \sqrt{S \times (AR)} = \sqrt{149 \times 6} = 30$ ft.

3. *Power Required*
   Thrust horsepower required:

   $$thp_{req} = \frac{1}{550} \times \left[ C_{D_p} \times \frac{\varrho}{2} \times S \times V^3 + \frac{2}{\pi \times \varrho \times e} \times \left(\frac{W}{b}\right)^2 \times \frac{1}{V} \right]$$

   For this equation, all terms must be in units of feet, pounds, and seconds; hence, 150 knots is converted to 253 ft/sec by multiplying by 1.69, the conversion factor. Then:

   $$thp_{req} = \frac{1}{550} \times \left[ 0.025 \times \left(\frac{0.001927}{2}\right) \times 149 \times \right.$$
   $$\left. (253)^3 + \frac{2}{\pi \times 0.001927 \times 0.6} \times \left(\frac{2533}{30}\right)^2 \times \frac{1}{253} \right] = 133.9 \text{ hp}$$

   Brake horsepower required:

   $$bhp_{req} = \frac{133.9}{0.8} = 167.3 \text{ hp}$$

Rated brake horsepower:

$$\text{bhp}_{\text{rated}} = \frac{167.5}{0.75} = 223.2 \text{ hp}$$

4. Range
   Total fuel available:

$$\text{Tot. fuel} = \frac{W_f}{W_G} \times W_G = 0.15 \times 2533 = 380 \text{ lbs.}$$

Range:

$$\text{Range} = \frac{(\text{Tot. fuel}) \times V_{\text{cr.}}}{(\text{SFC}) \times (\text{bhp}_{\text{req}})} = \frac{380 \text{ lbs} \times 150 \text{ kts}}{(0.5 \text{ lbs/hr/bhp}) \times 167.3 \text{ bhp}} = 681 \text{ n. mi.}$$

At the end of the first estimating cycle, we note that the desired range of 800 miles was not quite achieved. The amount of fuel actually required is determined as follows:

$$\text{Tot. fuel} = \frac{\text{Range} \times \text{SFC} \times \text{bhp}_{\text{req}}}{V_{\text{cr}}} = \frac{800 \times 0.5 \times 167.3}{150} = 446 \text{ lbs}$$

A new weight estimate can now be made using this figure for fuel weight (rather than $0.15 \, W_G$), and the entire process repeated. The new weight, of course, gives more power required, resulting in greater fuel consumption, so that several iterations are necessary to make everything agree.

In this particular case, another option might be considered. The range of 681 miles was arrived at by cruising at 75 percent power. If the designer could tolerate a slower pace, greater range could be achieved at a lower power setting. For example, 55 percent of the rated power of 223 hp would be 122.6 bhp, or, assuming the same 0.8 propeller efficiency, 98 thp. If the $\text{thp}_{\text{req}}$ were calculated for various values of V and plotted, the curve would appear as shown in Fig. 8-16. We see that to achieve a thp of 98, the airplane would have to cruise at 128 knots. At this speed and power setting the range would be as follows:

$$\text{Range} = \frac{(\text{Tot. fuel}) \times V_{\text{cr.}}}{(\text{SFC}) \times (\text{bhp}_{\text{req}})} = \frac{380 \times 128}{0.5 \times 122.6} = 793 \text{ miles}$$

This figure is very close to our goal of 800 and would probably suffice for a first estimation. Note that Fig. 8-16 also confirms our original determination of 133.9 hp required for 150 knots at 75 percent power.

Thus, with only a few quick calculations, we have a rough idea of the major parameters that size our airplane. Specifically, we have determined the following:

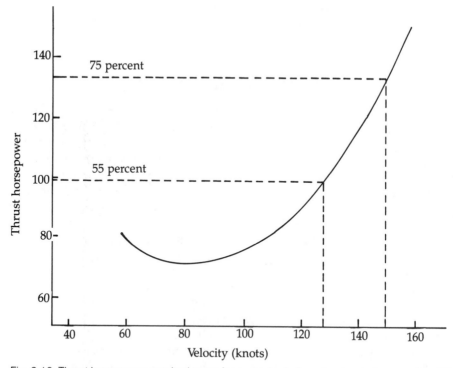

Fig. 8-16. Thrust horsepower required curve for example design, showing cruise speeds at 75 percent and 55 percent power.

| | |
|---|---|
| Gross weight: | 2,533 lbs |
| Wing area: | 149 sq ft |
| Wing span: | 30 ft |
| Brake horsepower: | 223 |
| Fuel required: | 380 lbs |

Notice that we do not even know what the airplane will look like at this point, other than that it will be a low-wing configuration, and that it must be fairly clean to have a low drag coefficient.

The tail for this first design estimation could be sized using typical tail volume coefficients, as described in this chapter's section on tail design; however, the design is not complete at this stage. It is really only the beginning. A detailed weight estimate is then done by considering the weight of each component of the airplane and totaling them. Drag is estimated more closely by drag breakdown methods, or by wind tunnel tests of models. The tail is configured by detailed study of the stability and control characteristics. All of these procedures are very complex and well beyond the scope of this book. With these more accurate values, the iteration process is continued until the design evolves into one with specifications met, or until the best compromise among them has been arrived at.

# COMPUTER-AIDED DESIGN

Computer-aided design is much touted these days by manufacturers of everything from television sets to baby bottles. Indeed, we have come to believe that if something is not computer designed, it must be old-fashioned. It is only natural that the computer would be employed in the design of something so complex and so precise as an airplane. But, there is nothing magical about a computer; it does not know any more about designing an airplane than engineers knew in the past. The computer is only a tool to perform voluminous calculations rapidly. The theoretical relationships that are programmed into it are the same ones that have been known for decades, even centuries, in some cases.

Aerodynamic engineers use the computer to model flow patterns. Classical fluid flow relationships are modeled by assuming many discrete points on a surface, rather than viewing the surface as a continuous shape. Thin airfoils and wings can be modeled to simulate their lifting characteristics by assuming that they are made up of a number of vortices that simply exist without any surrounding structure. They are required to behave in a certain way by imposing restrictions on the resulting flow, such as being tangent to the resulting surface.

The more points that are employed, the more accurate will be the resulting solution. Of course, each point is described by a mathematical equation, and all of these equations must be solved simultaneously to yield compatible results. Solution of more than just a very few points is a huge task to perform by hand. With many points, the solution could take years, if it is possible at all; however, the rapid-calculating capability of the computer enables such solutions in a few minutes, or even seconds, in some cases. The computer can also repeat the solution process many times to keep changing one parameter slightly until a desired result is achieved: iteration as previously explained.

Three-dimensional bodies can be modeled as a collection of flat panels, and fluid flow relations applied to each of these panels. Such techniques are referred to as *panel methods*. When combined with boundary-layer programs, they can be used to minimize the drag of various configurations. Again, the number of panels used, and their arrangement, determines the accuracy of the solution. The requirement of flow tangency is the restriction placed on the resulting flow. There are many different computer programs for such purposes. Selecting the proper one, and arranging the best panel distribution is often a challenging task for the engineer; however, once determined the optimum shape can be generated relatively quickly by the computer.

Figure 8-17 shows the paneling of a Beechcraft Staggerwing for computer analysis. This configuration, of course, was for a recent study; the Staggerwing was designed long before the computer was even dreamed of as a standard tool

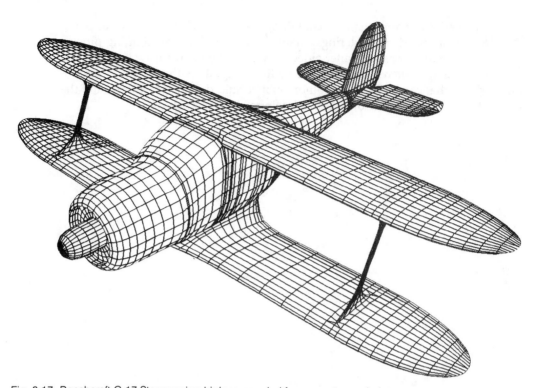

Fig. 8-17. Beechcraft G-17 Staggerwing biplane paneled for computer analysis. <small>David Lednicer, Analytical Methods, Inc.</small>

for aircraft design. Today, though, such techniques avoid time-consuming wind-tunnel tests of many alterations of a basic shape.

Structural engineers also enlist the aid of the computer in determining the loads and stresses in various structural elements of the aircraft. For many years there were analytic techniques to predict the stress in certain members and the size to resist this stress; however, many structural components are very complex, and these techniques could only approximate the actual stresses, and not in all of the elements; hence, many parts of the structure ended up being redundant or over-designed, meaning excess weight.

Nowadays, techniques known as *finite element methods* simulate large structural components with distinct, small elements, and their connections, known as *nodes*. By developing a series of mathematical equations to represent each element, and solving them simultaneously in a large array, a much more exact stress determination can be made for each part of the component. The result is a much more efficient, lightweight structure. Only modern computers are capable of the vast amount of computational work that must be accomplished to solve such problems.

Modern aircraft design teams also use the computer to generate drawings. The major advantage to this approach is the ease of making future changes that very frequently have to be done in the aircraft design business. Computers can also produce very neat and clear renderings. Many people think of this drafting process done by the computer as computer-aided design, and it is, in a way; however, the real contribution to aircraft design made by the computer is the optimum aerodynamic shape that can be made by computational fluid dynamic methods, and the efficient structure that is made possible by finite element analysis.

# 9

# Aerodynamic Testing

ONCE A DESIGN IS FINALIZED, the accuracy and wisdom of the design choices should be verified. One obvious way of going about this task is to construct a prototype, or first full-scale model, and actually fly it. The performance and handling qualities can then be evaluated firsthand and compared with what the designer predicted. Such testing, of course, cannot be done until the design is completed. It is highly desirable to have some test input into the design process before the configuration is "frozen," or put into final form. Testing in this phase must be performed with the models in the laboratory.

Hence, there are two primary types of aerodynamic testing of aircraft: *flight testing* and *laboratory testing*. Laboratory testing usually implies the use of a wind tunnel in which to conduct the tests, although not always. Wind tunnels are relied on to a great extent in model aerodynamic work. They are used by organizations such as NASA to perform basic research. This means the development of basic aerodynamic shapes, such as airfoils, inlet contours, or cowling shapes to perform most efficiently. Aircraft design companies usually use wind tunnels to test their particular designs. Sometimes this means the determination of the suitability of basic developments to their specific configurations.

More commonly, however, design organizations will use wind tunnels to make a better determination of the drag of a certain part of the design, or to test the stability effects of different wing-body-tail combinations. Sometimes a complete airplane model is tested or sometimes individual components. Because flight testing is a more obvious process, and wind tunnels are less understood, the bulk of this chapter will deal with wind tunnels and wind tunnel testing.

## HISTORY OF WIND TUNNELS

The first attempt at performing aerodynamic testing with models is credited to Sir George Cayley. He employed a whirling arm that moved the model through the air, rather than a wind tunnel. Langley also used such a device to aid in the development of his aerodromes.

The basic principle involved in the wind tunnel is to hold the model in a fixed position and force air to flow over it. This idea in testing was first suggested by Leonardo da Vinci. He postulated that the forces created were the same whether a body moved through the air or whether the air moved at the same velocity over a stationary body. Not everyone accepted this idea, however. Octave Chanute held that there was a difference in these two situations and debated at length over this point with the Wright brothers, who did believe it. Of course, we have since proven that da Vinci's theory was, indeed, correct.

The first wind tunnel recorded in history was developed by Francis Wenham in England in 1871. It was a very simple box with air blown through it by a steam-powered fan and was developed to measure lift and drag on some basic aerodynamic shapes. The second known tunnel was built by another Englishman, Horatio Phillips, about 1884. He conducted some very early experiments with airfoils. There is record of a few other tunnels being built in Europe and in Russia late in the 19th century.

The first American wind tunnel was built at the Massachusetts Institute of Technology in 1896. The wind tunnel was hence a relatively new and rare device when Wilbur and Orville constructed theirs in 1901. They had concluded from their early glider experiments that much early aerodynamic data was not entirely correct. Despite the contrary advice of Chanute, they constructed the tunnel shown in Fig. 9-1. It was powered by a gasoline engine driving a two-blade fan. They designed and built a special balance to measure lift and drag on airfoil shapes to be tested with this tunnel. The balance and several of the airfoil models are shown on top of the tunnel in Fig. 9-1. They tested several hundred different shapes and determined that the data obtained would actually enable them to predict the performance of an aircraft before it is built. One cannot help but marvel at the wisdom and ingenuity of these famous pioneers. It is not very often recognized that their contribution to the development of the science of aeronautics is almost as great as the development of the airplane itself. The wind tunnel and their clever use of it was another distinct step that led to the successful first flight.

Wind tunnels began to appear in Europe during the first decade of the 20th century as pioneers in aerodynamics developed the basic theories of that science. The next major step in wind tunnel development in the United States had to wait for the impetus of World War I. The National Advisory Committee for Aeronautics (NACA) was established in 1915, and in 1917 they began construction of the first laboratory facilities at Langley Field in Hampton, Virginia. The first wind tunnel at Langley was begun in 1919 and completed a year later. It was patterned after those developed by Gustave Eiffel in Paris. The Langley tunnel had a 5-foot-diameter test section and could develop speeds up to 120 mph. This tunnel was so successful that six months after it began operation a second tunnel was authorized and completed in 1922. This was a variable density tunnel (Fig. 9-2) capable of operating at high Reynolds numbers at relatively low speeds. It made possible the extensive development of airfoil data upon which much modern airplane design is still based today.

A 20-foot diameter tunnel was completed at Langley in 1927. Originally conceived for the testing of full-scale propellers, it proved to be quite useful in other

Fig. 9-1. Wright Brothers' wind tunnel.

aerodynamic developments using full-scale airplanes, as shown in Fig. 9-3. The success of these tests paved the way for the largest tunnel at Langley, which was completed in 1931. This was a full-scale tunnel with an oval test section 30 feet by 60 feet. It employed two fans each driven by 4,000 horsepower and could achieve a speed of 118 mph (Fig. 9-4). Eventually, an even larger tunnel was built at Ames Research Center in California with a test section 40 feet by 80 feet, the largest in the world.

## WIND TUNNEL DESIGN

The Wright brothers' tunnel was actually an *open-circuit* or *straight through* type. The Eiffel tunnel and the first NACA tunnel were also open circuit wind tunnels. This is a simple type of wind tunnel to construct and is very efficient. Many are still in use today, particularly in schools or in small research facilities. Figure 9-5 shows such a tunnel built by Aerolab of Laurel, Maryland, which is very popular in aeronautical school laboratories. Many of the illustrations of wind tunnel tests throughout the book were made with one of these tunnels.

Figure 9-6 shows a cutaway drawing of a typical open-circuit tunnel. The fan

Fig. 9-2. NACA variable density wind tunnel.

is located at the exhaust end to prevent the swirl of the slipstream from hitting the model. Straightening vanes are also used on the intake end to provide a smooth, uniform flow of air. The tunnel necks down to a smaller area through the test section to provide maximum velocity. The overall tunnel shape resembles that of a venturi tube.

Small open-circuit wind tunnels are usually operated inside of buildings. Larger tunnels must be open to the outside air for sufficient air supply; therefore, they are susceptible to turbulent, gusty air and to dust, precipitation, or other foreign material in the atmosphere. Open-circuit tunnels are also very noisy and surrounded by high wind currents, making them difficult to work with.

## Closed-circuit tunnels

Larger wind tunnels are nearly always made in the *closed-circuit* or *return-type* configuration, such as illustrated in Fig. 9-7. Air is accelerated by the fan and flows through the tunnel. Turning vanes are installed to guide the flow around the corners. The tunnel widens into a large settling chamber to decelerate the air to a relatively low velocity. This prevents the build-up of a large boundary layer along the walls, which would occur if the entire circuit were at a near constant velocity.

The flow then is contracted through an entrance cone, or nozzle, which causes the velocity to increase to a maximum as it enters the test section. A contraction ratio (area of the settling chamber to area of the test section) of at least 4 is

Fig. 9-3. NACA propeller research tunnel.

necessary, with larger ratios desirable. Figure 9-8 shows the contraction nozzle and entrance to the test section of the low-speed tunnel at Penn State University's aeronautical laboratory. This tunnel has a test section 4 by 5 feet in cross section, with a settling chamber 12 by 12 feet, yielding a contraction ratio of 7.2. The model is mounted and all tests conducted in the test section. Immediately after the test section a diffusion section, or gradually widening chamber, decelerates the air to a lower velocity before it again reaches the fan.

The wind tunnel described is really a *single-return* type. There are also *double-return* tunnels that divide the flow downstream of the test section and run it through two circuits, each with a driving fan. The flow then joins as it enters the settling chamber and proceeds through a single test section.

## Annular tunnels

An extension of the double-return concept is the *annular-return*, in which a return passage is located around the entire circumference of the wind tunnel outside of the test section. The annular-return design was employed in the NACA variable density tunnel. This configuration lent itself very well to a pressure

Fig. 9-4. NACA Langley full-scale wind tunnel.

Fig. 9-5. Aerolab open circuit demonstration wind tunnel.

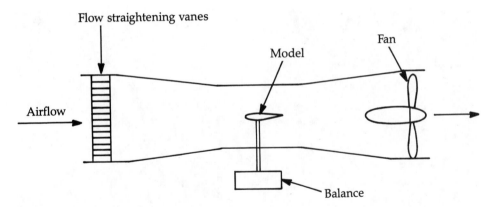

Fig. 9-6. Diagram of an open circuit wind tunnel.

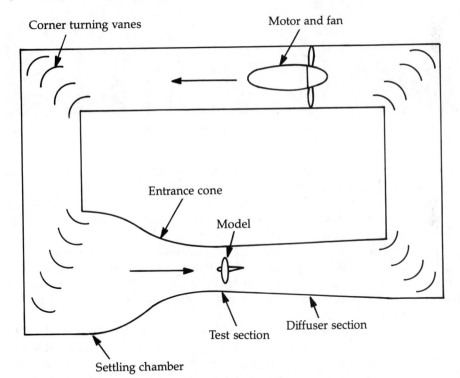

Fig. 9-7. Diagram of a closed circuit wind tunnel.

chamber design and was selected for that particular purpose. The test section location is difficult to access and view; thus, this design is rather rare.

## Spin tunnels

A special type of annular tunnel is employed at Langley Research Center for spin research. This tunnel, shown in Fig. 9-9, is mounted vertically with the fan

Fig. 9-8. Contraction nozzle of the Aerospace Engineering Department wind tunnel at Penn State University.

drawing air upward. Models are introduced into the test section in free-flight condition. The velocity is adjusted so that the vertical flow just balances out the gravitational pull on the model and it remains suspended. The test section is also designed to have a slightly lower pressure at the center than at the walls. This tends to hold the model in the center of the test section.

If the model is injected in a configuration causing it to spin, the spinning will continue as the model remains suspended. Its spin rate, attitude, and other characteristics can then be photographed and studied. Other tests are sometimes conducted in spin tunnels, such as parachute performance and aircraft tumbling phenomena.

Spin tunnels are rather rare, with only a few existing in the world. In the 1970s a resurgence of interest in spin research developed. NASA performed a considerable amount of study on spin characteristics of light airplanes. These tests involved radio-controlled model tests and full-scale flight tests, as well as spin tunnel investigation.

## TYPES OF WIND TUNNEL TESTS

To those who are not familiar with the workings of aeronautical labs, the actual methods of wind tunnel testing could appear somewhat mysterious. Sticking a model into a big tunnel of air and coming up with specific numbers on its lift and drag properties might seem more like the work of a sorcerer than an engineer.

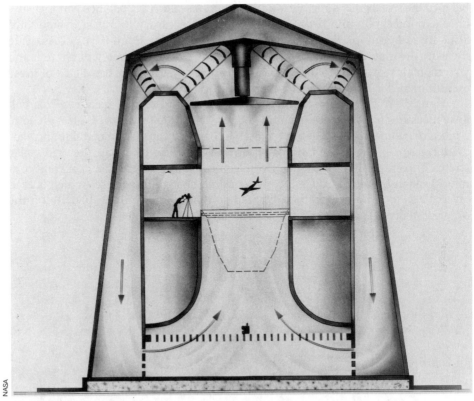

Fig. 9-9. NASA Langley 20-foot spin tunnel.

The actual tests are straightforward and simple—at least in their general approach. Correcting for inaccuracies of small models and accounting for non-uniform flow and imprecise instrumentation can pose more complexities.

Although a variety of special tests can be performed in wind tunnels, most testing is done to perform one of three basic tasks:

- Force measurement
- Pressure measurement
- Flow pattern study

## Force Tests

The original reason for inventing the wind tunnel was to provide a means to determine the lift and drag on airfoil shapes. Wenham, Phillips, and the Wright brothers all used wind tunnels for this specific purpose. Lift and drag are forces, so we term such testing *force measurement*. Force measurement requires some sort of balance or other device to determine the number of pounds (or whatever unit) of force exerted.

The small Aerolab wind tunnel shown in Fig. 9-5 uses a simple mechanical balance that reads out directly by a pointer on a scale. This balance can measure only two forces: lift and drag. Quite often it is necessary to measure side force also. In addition—particularly when doing stability testing—it is necessary to measure the moments about all three axes of the airplane, namely pitch, roll, and yaw moments. A more complete balance is one that can measure all six of these quantities and is termed, appropriately, a *six-component balance*.

Figure 9-10 shows a typical six-component wind tunnel balance. This balance measures forces by use of strain gauges. These are tiny sets of wires glued to a piece of metal subjected to the forces. As the force causes the metal to increase or decrease in length (depending upon the direction of load), the wires also stretch or contract. This changes their electrical resistance, and so a current running through them will cause a corresponding change in voltage, which is read on a voltmeter. The balance is calibrated so that so many volts (usually on the order of millivolts) represents so many pounds of force.

Fig. 9-10. Six-component pyramidal balance for measurement of forces and moments in the wind tunnel.

This type of balance is very accurate and can be made to be nearly free of cross-coupling. One of the major problems in balance design is to make it so that each force and moment can be measured "purely," or without any input from a force or moment in another direction. Because all measurements are taken from a common mounting point (usually one or two posts connecting the model to the balance), this can be a difficult task. Most balances will have some coupling effects and the measurement of one force—say, lift—requires reading several channels on the balance readout. A calibration equation or set of curves is then provided to translate these readings into force or moment.

A typical example of force measurement is the determination of lift coefficient on a certain airfoil shape. A wing of this airfoil contour is placed in the wind tunnel and connected to a balance. The tunnel is run at a certain airspeed with the wing at a given angle of attack, and the lift force measured. The lift coefficient for this condition is determined by dividing the measured lift by the product of dynamic pressure ($1/2 \times \varrho \times V^2$) and the wing area.

$$C_L = \frac{L}{q \times S}$$

The density is determined from measurement of the temperature and pressure of the air in the tunnel. All tunnels will have a device for measuring airspeed. Usually a pitot-static tube is used, just like on an airplane. The two-dimensional airfoil properties can be approximated if the wing model goes all the way to (but not quite touching) the tunnel wall. In this way the tip vortex is eliminated and the wing behaves like a two-dimensional airfoil.

## Pressure tests

A force measured as just described gives the overall lift force on the model. It cannot reveal how that force is distributed over the surface of the wing. Many times researchers also want to know the distribution of pressure in the wake or perhaps in the boundary layer. For such measurements researchers insert tiny tubes into the model surface or the airstream (depending on the exact test) and connect them to a pressure measuring device.

The old standby in pressure measuring devices is the *liquid manometer*. To illustrate the use of a manometer in measuring the chordwise pressure distribution over an airfoil, consider the model shown in Fig. 9-11. Tiny holes, *pressure taps* as they are called, are drilled in the model top surface. Tiny brass tubes are inserted just flush with the surface and connected through flexible tubing to a manometer. This is a series of tubes all standing in a common well of liquid (water, in this case). Initially, all of the tubes have the same water level due to gravity acting equally on all of them.

Figure 9-12 shows the wing in a running wind tunnel. The lowered pressure over the wing surface reduces the pressure in the manometer tubes and draws the water up to a higher level. The lower the pressure becomes, the higher the water level goes. By measuring the difference between the height of the water in a given tube and one open to the ambient air, the relative pressure difference can be calculated. Notice that the height of each tube is different, indicating a different static pressure at each point on the wing surface. If the wing is adjusted to a higher angle of attack, the pressure will be lowered and its distributions shifted more toward the leading edge, as shown in Fig. 9-13.

The airspeed in the tunnel is often measured by connecting the pitot and static tubes to two of the manometer tubes. In this case, the total (or pitot) pres-

Fig. 9-11. Model wing with pressure taps for wind tunnel pressure tests.

Fig. 9-12. Pressure distribution as measured over model wing by use of a manometer.

Fig. 9-13. Altered pressure distribution of model wing as measured at high angle of attack.

sure will be higher than atmospheric, and will push the water down in the tube. The static pressure tap will draw it up, and the difference between these two will indicate the dynamic pressure, from which a velocity can be calculated. Studies of the free stream or wake pressure distributions involve a series of pitot-like tubes pointed into the airstream. Such a device is referred to as a *wake rake*.

## Flow patterns

The way that streamlines of air flow over a body has a great deal to do with that body's aerodynamic properties; therefore, it is very desirable to be able to see just what the flow pattern looks like. Unfortunately, air is invisible, so that some means must be devised to visualize the flow.

One of the tried and true methods of flow visualization that has been around for many years is *tufting*. Tufting employs the attachment of small tufts of yarn (or thread, depending on model size) to the surface of the model. They are attached by small strips of thin tape, such as cellophane tape. When mounted in a running tunnel, the tufts will indicate the direction of the flow in the vicinity of the tufts. Furthermore, they will show if the flow is attached, in which case they will point directly downstream, or separated to form a wake. If the tufts are in a wake, they will flutter around and point in random directions, often into the incoming free airstream. Figure 9-14 shows a tufted model.

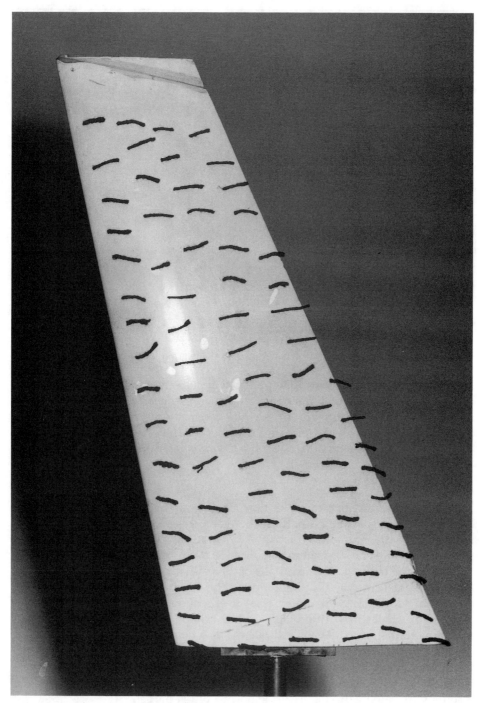

Fig. 9-14. Model of a forward-swept wing, tufted for wind tunnel flow pattern study.

If the flow somewhat above the actual surface is to be studied, tufts can be mounted atop small posts such as straight pins.

Tufts can also be used to show the flow pattern in a wake downstream of the model. Wingtip vortices present a typical subject to be studied here. In this case, a *tuft grid* is constructed. This is a large-mesh screen with tufts tied at each intersection of the wires. This is installed at the desired point downstream in the tunnel. In either way that tufts are employed, their resulting indications can be photographed and used to provide valuable information about the flow.

Another old reliable method of making the flow visible is the use of smoke. Smoke is generated by burning oil or some similar substance and then injecting the smoke into the airstream. The visible smoke patterns can be observed or photographed just as tuft patterns can be. Smoke tends to obscure parts of a three-dimensional model, so it is most often used to visualize two-dimensional models in narrow tunnels created especially for this purpose.

Smoke is awkward to handle, however. If used in an indoor tunnel it will quickly fill the laboratory, making it difficult for the operators to breathe. This can be a hazardous situation if prolonged. Furthermore, oil-fired smoke contains oil particles and it can make a mess of both the model and the interior of the tunnel. Figure 9-15 shows a smoke tunnel test in progress.

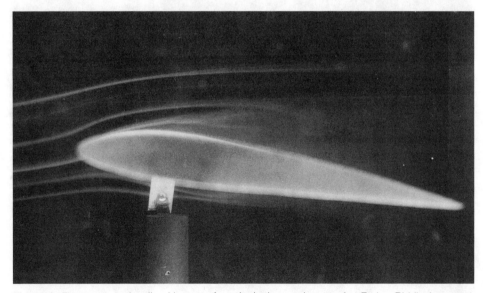

Fig. 9-15. Flow pattern visualized by use of smoke in the smoke tunnel at Embry-Riddle Aeronautical University.

Oil flow techniques are also used to study the flow pattern on models. The surface is sprayed with a light coat of oil prior to the test. Sometimes the oil is darkened with lampblack or a similar substance. The direction of the streamlines will then be indicated by the way in which the oil flows over the surfaces. Separation of the flow and transition from laminar to turbulent boundary layer can be indicated by this technique. The visualization can be enhanced by use of a fluorescent oil viewed under ultraviolet light.

# HIGH-SPEED WIND TUNNELS

As flight speeds crept higher and higher during the 1930s, wind tunnels were developed to provide equivalent velocities. A small high-speed tunnel was built at Langley even before 1930. In 1936, an eight-foot diameter tunnel was built that could achieve 575 mph. By 1945, its speed capability had been increased to Mach 1 (approximately 760 mph). The most powerful high-speed tunnel of this era was the 16-foot tunnel constructed at Ames just prior to World War II. Its 27,000 horsepower electric motor drove it at near sonic velocity and provided the testing capability for most of the fighters of the time. Various airplanes from P-40s to P-51s were tested in this tunnel. It is shown in Fig. 9-16. These tunnels provided speeds in the high subsonic range, but just barely reached the speed of sound, if at all.

Fig. 9-16. NASA Ames 16-foot tunnel completed just before Pearl Harbor Day in 1941.

## Supersonic tunnels

The first practical supersonic wind tunnel was developed around 1935 by Adolf Busemann, in Germany, the scientist also credited with the development of the swept-wing concept. His tunnel was the example from which practically all supersonic wind tunnels were developed. The first American supersonic tunnel was designed by the famous aerodynamicist Theodore von Karman at Cal Tech in 1944. These early supersonic tunnels were of the "blow down" type. A large tank of compressed air was pumped up and then released to blow down through the tunnel. The air goes through the throat, where the flow is exactly Mach 1. It is then expanded into a test section where the flow becomes greater than Mach 1. The actual Mach number in the test section depends upon the ratio of its area to the throat area, and is controlled by adjusting the area of the throat. These tunnels are relatively simple to construct in small scale, but are limited to only a few minutes—or seconds—of run time.

By the 1950s, continuously running supersonic tunnels were developed. One of the largest was built at the Arnold Engineering Development Center in Tennessee, which had a test section of 16 feet square. Small tunnels were also built during this period that could achieve speeds in the hypersonic range, greater than Mach 5.

One of the most difficult conditions to create is flow right in the area of Mach 1. Flow in this transonic region is terribly unstable and very hard to maintain continuously. The latest major wind tunnel facility to be developed is NASA's National Transonic Facility at Langley, designed for studies in this area. It was completed in 1983 and consists of a 16-foot tunnel that can achieve Reynolds numbers of 120 million at Mach 1. Such flow is in the cryogenic region, with temperatures on the order of $-250\,°F$, and requires the use of liquid nitrogen. This facility is shown under construction in Fig. 9-17. A schematic is also shown in Fig. 9-18.

NASA

Fig. 9-17. The National Transonic Facility under construction at NASA Langley in 1980.

## Shock wave visualization

One of the major items that researchers must deal with in supersonic flow is the shock wave, which has a great effect on high-speed aerodynamics. Visualiz-

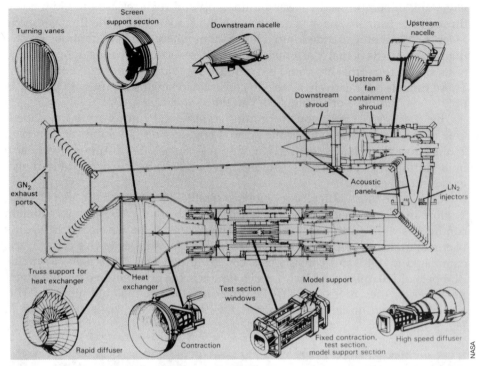

Fig. 9-18. Diagram of the National Transonic Facility.

ing the shock wave pattern is thus of great interest here. Shock waves are normally invisible, just like any other airflow pattern. They are made visible by various devices that take advantage of the light refraction effect of the much denser air in the shock wave. The most common method of utilizing this principle is the *Schlieren method*.

In the Schlieren system, a high intensity beam of light is projected through the shock wave in the wind tunnel test section. The shock wave deflects the rays that pass through it, just as a glass of water deflects the rays passing through the water and appears to offset the scene to the viewer. These deflected rays are projected over a carefully adjusted knife edge, which blocks them out. The entire view of the test section is then projected onto a screen on which the blocked rays (the ones that passed through the shock wave) appear darker than the surrounding air. Figure 9-19 shows a diagram of a Schlieren apparatus. Figure 7-3 was made by use of this technique.

## Wind tunnel testing problems

Early wind tunnel experimenters worried about the validity of blowing air over a stationary model to simulate actual flight conditions. This concern has long since been dispelled. Wind tunnels are not without problems, however. They are plagued with many inaccuracies and cannot duplicate exactly the conditions encountered in free flight.

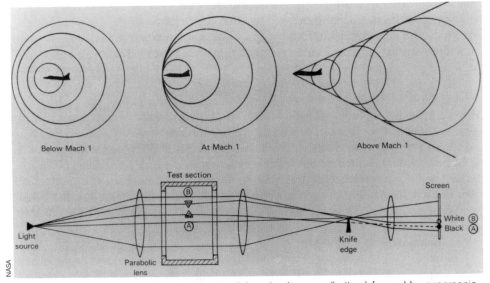

Fig. 9-19. A Schlieren optical system for visualizing shock waves (bottom) formed by supersonic flow as shown at the top.

## Wall effects

The walls of the wind tunnel represent artificial boundaries, which do not exist in actual flight. They therefore impose certain constraints on the flow. For example, the model itself takes up space in the test section. A portion of the test section area is thus blocked and the actual area for the flow is reduced. By the law of continuity, the flow would have to speed up around the model from the free airstream velocity. In actual flight, there is no lateral constraint to the flow and it would expand around the model.

Upwash and downwash effects also extend some distance from the model. If the walls (including floor and ceiling) interfere with such flow, the result is the same as an endplate or ground effect. The lift and drag measured would then not represent the true values found in the free air.

One way of eliminating some of these effects is to have an open test section with no walls. This cures some of the problems, but creates others. An obvious answer is to make the model so small that the walls are far enough away to have negligible effect. The problem here is that small models will have small forces, making them difficult to measure accurately. Another major problem exists with small models and is made even worse by reducing the model size. This is known as *scale effect*, and is due to the difference in Reynolds numbers between the model and the full-scale airplane.

## Scale effect

Reynolds numbers are discussed in chapter 3 and have to do with the relative viscosity of the air, which affects the boundary layer. The Reynolds number is the velocity of the air times the length of the surface over which it flows divided by a viscosity coefficient. At standard sea level this coefficient is 0.00158 ft²/sec. Dividing by this is the same as multiplying by 6,329. A Reynolds number for a wing is thus the velocity (in ft/sec) times the chord (in feet) times 6,329.

$$R_e = V \times c \times 6,329$$

For example, a wing in a 100 mph wind (146.7 ft/sec) with a 5-foot chord would have a Reynolds number of approximately 4.6 million.

Boundary layers normally start out as laminar boundary layers and transition to turbulent at a certain Reynolds number, depending upon the amount of turbulence present. For average flight conditions, this is somewhere on the order of 300,000. This Reynolds number on the wing that was most recently described would be reached at approximately 4 inches back from the leading edge. At this point the boundary layer would become turbulent and most of the wing would be in such flow.

Now suppose we wanted to simulate this wing in a wind tunnel with a model wing of 4 inches in chord length. The entire wing would be in a laminar flow area. Because there is a significant difference between the drag of a laminar boundary layer and a turbulent boundary layer, the model would show considerably less drag than the real wing, even when the measured drag is scaled up. Overcoming scale effect is a problem.

Making the model larger would be useless unless we could reach the full-scale size. Running the tunnel faster might not be possible, and even if it were, we would probably have to go so fast that we would encounter Mach effects, again invalidating our test data. One probable solution is pressurizing the wind tunnel as NACA did in its variable-density tunnel; however, this is a complicated and expensive proposition. The usual solution resorted to is to artificially induce the boundary layer to go turbulent by placing some grit or other substance on the model surface. This does not reproduce the Reynolds number of the full-scale, but it does represent similar boundary layer distribution. Model tests must be carefully interpreted and used cautiously. They give a good indication, but not always an exact prediction, of the behavior of the full-scale aircraft.

## FLIGHT TESTING

The real proof of the design is in the actual flying, properly called *flight testing*. Once a design is finalized, a prototype is constructed for this purpose. The prototype is handmade and might differ in certain respects from the forthcoming production model.

Flight testing is performed in several different phases. The first phase is a series of shakedown tests, in which the basic flying qualities are determined. Items such as engine cooling, vibration, and unusual control response are carefully monitored to ensure that the remaining tests can be performed safely. If the airplane does not pass these tests, the design goes back to engineering for corrective action. If the airplane does all right, testing moves into the next phase.

The next phase is a determination of the airplane's performance. A series of tests are flown in which the various items of performance are carefully measured, usually with the aid of special instrumentation. An exact determination can then be made of such items as top speed, cruise speed, range, rate of climb, takeoff distance, and landing distance. This procedure will verify how well the airplane meets the required specifications, and hence how well the design has met the designer's predictions. (A more detailed description of flight test procedures and how they can be performed on light airplanes is found in my book, *Performance Flight Testing*, published by TAB [*see* bibliography].)

The final phase in the flight testing process is a thorough examination of stability and controllability. A more detailed study is made here than in the initial shakedown tests. Quantitative measurements of the exact degree of stability are determined in this phase, whereas only basic handling qualities are evaluated in the first tests. Test pilots are relied upon in the initial testing phases for their opinions of the airplane's reactions; however, test engineers usually perform the actual measurement in the latter phases, with the pilot's job being to accurately fly at constant, given conditions.

In wind tunnel testing, forces (such as lift and drag) are measured and the airplane's—or at least the model's—performance is then predicted from calculation. In-flight testing, the opposite is done. Performance is measured, and lift and drag are calculated by working back the other way in the applicable mathematical equations. Drag on the real airplane, for example, is determined in this fashion, because the airplane in flight cannot be attached to a balance, as are wind tunnel models. In either case, an actual drag measurement can be made. Flight testing of different configurations more accurately determines the configuration with least drag.

Pressure measurements can be made in flight by attaching pressure measuring devices to taps on the aircraft surface. This is quite similar to wind tunnel procedures. Probes can also be mounted away from the surface of the aircraft to examine the flow in the wake. Figure 9-20 shows an apparatus for mounting a probe to measure vortex velocities behind the wing of a Cessna 0-1 observation airplane. Flight test projects such as this are intended not to refine the specific design, but rather to perform basic research for better understanding of aerodynamic behavior. Airplanes employed in such study are often nothing more than flying test beds to enable tests on certain components to be conducted in flight rather than in the wind tunnel.

Flow visualization can also be performed on flight test aircraft by use of tufts, similar to procedures with wind tunnel models. Yarn tufts of several inches in length must be used for the full-scale aircraft, and are usually attached with fairly sturdy tape, such as masking tape. Flow patterns on portions of the airplane visible from the cabin can be observed from the test aircraft. Usually it is necessary to photograph the tuft-covered area from a chase airplane flying in formation with the test aircraft.

Tuft tests can reveal areas of poor aerodynamic design by the evidence of wakes resulting from separation of the flow from the surface. Such tests are often run to refine the design of a wing fillet or a fairing around a wheel or strut. Some-

Fig. 9-20. This apparatus is used to mount a probe to measure vortex velocities behind the wing of a Cessna 0-1 observation airplane.

times testing of this nature is done even after production models are being turned out. The results will be applied to future models, or even to entirely new model designs being contemplated. In this way, testing closes the loop in the design process and paves the way for improved efficiency in the next design.

The use of tufts in studying stall patterns on a wing is shown in Fig. 2-42. In Fig. 9-21, tufts are shown being used to indicate the flow over an extended flap. Note that many of the tufts are fluttering in the separated flow, indicating that a large wake, and hence, a lot of drag, is being produced. Of course, in the case of a flap, drag is often desirable for improved landing performance.

Special instrumentation is usually employed in professional flight testing, due to the inherent limitations of standard instrumentation. One of the major requirements in most flight testing is a very accurate measurement of airspeed. For this reason, one of the first pieces of special equipment added is usually a boom-mounted pitot-static probe. Figure 9-22 shows this temporary addition to a Piper Arrow. The boom places the probe far enough ahead of the wing to avoid interference from the flowfield around the wing, and enables accurate sensing of the free airstream. The probe is also mounted on a swivel, and has vanes to align the pitot tube exactly with the airstream. The probe is then connected to a specially-calibrated sensitive airspeed indicator, which can be seen mounted on top of the instrument panel just inside the windshield.

Figure 9-23 shows the boom in use as the airplane is being slowed into a stall. Note the pitot tube angled downward significantly from the longitudinal axis of

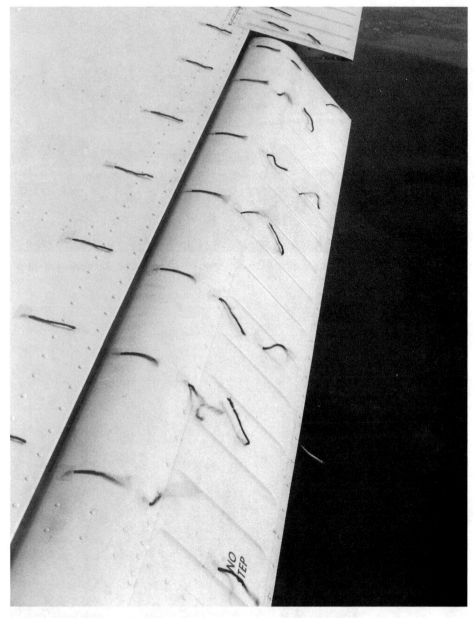

Fig. 9-21. Tuft studies of flow over an extended flap.

the airplane, which is the same as the alignment of the boom. This direction of the free airstream is due to the descending velocity of the aircraft in the stall approach, as well as the forward velocity. Remember that the standard pitot tube is fixed, and, hence, pointed in the direction of the boom.

Such special devices eliminate much of the error inherent in standard systems. Many of these errors are small, and the standard instrument systems are

Fig. 9-22. Piper Arrow II with boom-mounted swivel-head pitot-static probe and sensitive airspeed indicator.

Fig. 9-23. Boom-mounted pitot-static probe being used to measure airspeed in approach to a stall.

certainly accurate enough for normal operation; however, in measuring and comparing performance, as, for example with small drag-reduction devices, a few knots make a lot of difference.

# 10

# Modern Design Concepts

AIRCRAFT DESIGN HAS INDEED COME A LONG WAY in the nine decades since Orville first lifted off the sands of Kitty Hawk. The minimum speed of many airplanes is now well beyond the top speed of airplanes of the first decade of flight. The top speeds have increased by a factor of 50. Altitude capability is almost unlimited and is restricted by such things as pressurization, radiation exposure, and environmental concerns, rather than climb performance. Means have been devised to enable airplanes operating at these outer fringes of performance to still take off and land at reasonably short, established airports. Even personal light airplanes now outperform most airliners and many military aircraft of the 1930s.

## DESIGN EVOLUTION

Aerodynamic research and development have undoubtedly contributed significantly to the present state of the art in design; however, the world of aerodynamics has seen few really dramatic breakthroughs. Most aerodynamic improvement is the result of a gradual evolution in efficient aircraft configuration. Aerodynamic refinement has often followed more revolutionary developments in other branches of aeronautics. Propulsion is an area that stands out in this respect.

One of the major reasons for the delay in achieving powered flight in the beginning was the lack of a sufficiently powerful and lightweight engine. Solving this aspect of the problem was a large contributor to the Wrights' success. The development in reciprocating engine technology during World War I was largely responsible for the greatly improved performance of the last generation of World War I airplanes. The success of many record-setting flights soon after the war can also be attributed to this fact. The Orteig prize for the first direct flight from New York to Paris was first offered in 1919, but it was not until after the Wright J-5 was developed in 1926 that Lindbergh's flight was really made possible the following year.

Another area that facilitated aerodynamic improvement was structural design. The conventional wood-and-fabric structure with external bracing predominated aircraft design through World War I. In addition to strength and weight problems associated with this type of structure, it simply did not lend itself to clean aerodynamic design. A significant step toward low-drag configuration was the development of the cantilever, or internally supported, wing. Such construction was pioneered by Hugo Junkers and Anthony Fokker in World War I fighters.

Another major development of this era was monocoque structure, in which the skin not only forms the contour, but also carries some of the structural loads. This concept was first employed in the wooden fuselage of the French Deperdussin racer prior to World War I. Junkers built the first all-metal aircraft during the war, but it relied on external corrugations for load-bearing qualities. Adolph Rohrbach actually developed the smooth-skin metal monocoque structure and applied it to wings as well as fuselages.

All of these developments in propulsion and structure laid the groundwork for major improvement in aerodynamics. In the latter part of the decade of the 1920s, the National Advisory Committee for Aeronautics (NACA) had established a good research facility at Langley Field, Virginia. A very significant development to come out of that facility was the NACA cowling. This was a cowling shape, developed to perfection in a wind tunnel, to reduce the high drag associated with radial engines of that era. It was first developed for the Wright J-5 and resulted in reducing the drag of that engine by 60 percent. When fitted to a Curtiss AT-5A trainer, the speed increased from 118 to 137 mph, equivalent to adding 83 hp to the Wright's rated 220 hp.

Another NACA finding of this period was that the optimum location for engine nacelles was to be faired right into the wing. This arrangement gave the lowest drag. Tests with wind tunnels also proved that landing gear was a major drag contributor and that gear retraction was a highly desirable characteristic, despite the added weight and complexity. The landing gear of the Sperry Messenger, for example, was shown to account for 40 percent of the total drag of the airplane. The drag reduction of wing fillets was also recognized about this time. Developed first by California Institute of Technology, they were later refined by the NACA.

If ever there was a time that could be called revolutionary in low-speed aerodynamic development, it would be this period of the late 1920s and early 1930s. Most of the innovations of this era were incorporated in the Boeing 247, often termed the "first modern airliner," and the Douglas DC-2 and DC-3 that followed shortly thereafter. Indeed, they set the trends in design that were to exist for several decades.

Much of the design concepts employed in the DC-2 and -3 had been developed by Jack Northrop. Prior to his involvement with Douglas (and with his own company), Northrop had been an engineer for Lockheed. He was primarily responsible for the development of the first American airplane to employ a cantilever wing and wooden monocoque structure: the Lockheed Vega. Later fitted with an NACA cowl and highly faired landing gear, this famous bird cruised at 160 mph. It was flown by nearly every famous record-setting aviator of the time:

Fig. 10-1. The Lockheed Vega *Winnie Mae* flown by Wiley Post.

Lindbergh, Amelia Earhart, Roscoe Turner, and Wiley Post (shown with his Vega, the *Winnie Mae*, in Fig. 10-1).

World War II was the setting for the development of the jet engine. This was truly a revolutionary item and was responsible for the single most notable jump in performance in aircraft history. The jet engine only made high-speed flight possible, however. It took a concentrated effort of study in a whole new realm of aerodynamics to bring it to reality. Shock waves and their effects had to be understood and dealt with. New airfoils, new nose and inlet shapes, and new wing shapes had to be developed by the aerodynamicist before the thrust of the jet engine could push airplanes near to, and above, the speed of sound.

Low-speed aerodynamics was almost totally neglected in the jet age and light aircraft design relied pretty much on prewar technology. One significant exception to this trend was, again, in engine technology. It received little press, with jets capturing all of the headlines, but the reciprocating engine emerged from World War II as a highly refined piece of machinery. After all, it powered more than 99 percent of the airplanes used in the war. And the piston engine was—and still is—the ideal powerplant for lightplanes. Manufacturers now had the technology to make it more efficient, lighter in weight, and more reliable than ever before. The flat, horizontally opposed type was quickly zeroed in on as the ideal arrangement for low- drag installations.

Most light airplane manufacturers used only very low-powered engines and applied them to tube-and-fabric airframes. The result was the very light sport airplane, which offered much in the way of fun flying but little in real utility. A few manufacturers turned out metal airplanes with moderate performance, but one clearly stood out. Beechcraft presented the world with the first truly modern lightplane—the Bonanza (Fig. 8-14).

Capable of 175 mph on only 185 horsepower, this airplane was the trendsetter for a generation of business and personal aircraft. It was "the DC-3 of lightplanes," introduced in 1947 and was as modern in appearance as if it had been developed 40 years later. It took another decade, though, before the world awoke to the potential of the light airplane for business and personal business use. Figure 10-2 shows the V-35 Bonanza, the last model with the V-tail. The most recent production models utilized a conventional vertical and horizontal tail combination.

Fig. 10-2. Beech Model V-35 Bonanza.

Another milestone in general aviation was the introduction of the light jet, more commonly referred to as a *business jet*. Although several other small jets existed at the time, credit for the success and popularity of this type of aircraft must be accorded to Bill Lear. The Learjet was truly a light jet, capable of operating from 2,300-foot runways, yet attaining a speed of 500 mph and a rate of climb of nearly 7,000 feet per minute when introduced in 1963 (Fig. 7-8). Lear borrowed high-speed technology from military and airline aircraft, but applied lightweight structural concepts that had been practiced with conventional lightplanes.

Piston engines still dominated the general aviation field due to their much lower cost. By the 1970s the fuel efficiency of the piston engine became a dominant factor. Aerodynamicists also contributed to fuel conservation. The concept of "drag cleanup" was long known in higher performance circles. Now it became popular with even the slowest trainers.

Drag cleanup is a process of reducing the drag by carefully improving the efficiency of airflow around minor, detailed parts of the airplane. Fairing in the juncture of struts or improving filleting at the wing and tail surface roots are typical examples. Figure 10-3A shows the fairing on a Cessna landing gear strut as

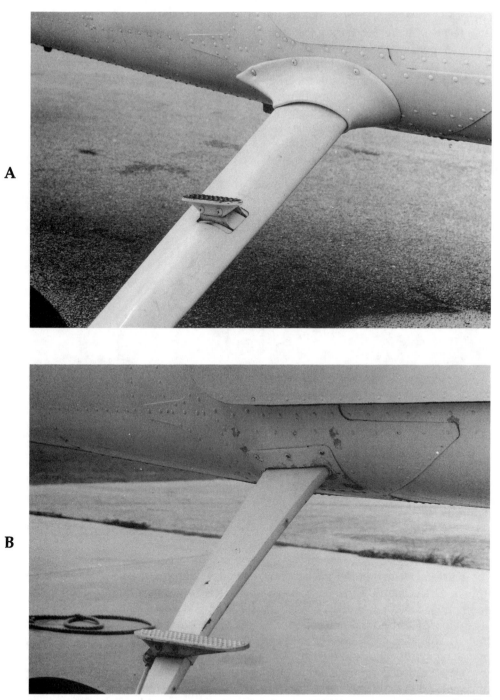

Fig. 10-3. Faired strut and strut junction of modern Cessna 172 (A) as compared to strut of earlier model (B).

compared to the unfaired earlier model in Fig. 10-3B. Entire landing gear struts also received increased fairing, as seen on the Piper Warrior shown in Fig. 10-4. Cowl shapes were also remolded. Even the performance of the efficient Mooney was vastly improved in the late 1970s by a new cowl design and other cleanup in the Model 201. The contrast of this shape is seen against an earlier Mooney cowl in Fig. 10-5. The Mooney, incidentally, was also rather innovative when first introduced in 1955. It represented an early attempt at low-profile aerodynamically clean, fuel-efficient design in the business/personal airplane class. It too was about 20 years ahead of its time.

Fig. 10-4. Highly refined fairing of wheel and strut on modern Piper Warrior.

Currently, the trend of gradual improvement in aerodynamic efficiency continues. Drag cleanup gains a few knots of airspeed or a fraction of a mile per gallon here and there. Collectively such efforts add up to a significant improvement. The Mooney 201 added more than 20 knots to the previous Mark 21, which, as noted, was already a very efficient design. Many companies now offer drag reducing modifications to older model aircraft. The modifications are not cheap, but some of them produce dramatic performance improvement, another testimony to the effectiveness of attention to aerodynamic details.

Strips to seal the gap between control surfaces and the fixed surfaces to which they are attached are becoming popular as one simple, low-cost drag modification. Their effectiveness depends on how much drag the gap created in the

Fig. 10-5. Improved aerodynamic shape of Mooney 201 compared with earlier Mooney Mark 21.

first place. Figure 10-6 shows the before and after situations with flap and aileron gap seals applied to a Piper Arrow II. Flap hinge fairings were also added to this airplane, and are apparent in the photo, too. Tests have shown such mods to be moderately effective in drag reduction, with cruise-speed increases of 3 or 4 knots; however, when combined with other equally effective mods, the total drag reduction could be significant.

The 1970s saw the return of attention by NASA, manufacturers, universities, and other researchers to low-speed aerodynamic improvement. As a result, a number of new design concepts emerged. Some of them were new applications of old ideas, but nevertheless, distinctly new configurations began to appear. Let us look at some of these concepts in more detail.

## CANARDS

One of the seemingly newer concepts in general configuration is the *canard*, which is a forward-mounted horizontal tail arrangement. This design appears in many drawings of futuristic or "dream" airplanes. The concept is not new and was employed by the Wright brothers in the very first airplane. Their choice of a canard was primarily for stall protection.

One of the leading proponents of canard designs is Burt Rutan, with his VariEze and its relatives (Fig. 10-7). These airplanes and their offspring have achieved remarkable performance with relatively low power. Much of this perfor-

Fig. 10-6. Piper Arrow wing in original condition (left), and with gap seals and flap hinge fairings (right).

mance is due to the very lightweight construction, low frontal area, and smooth surfaces, as well as to the canard arrangement.

The major appeal of the canard is that it provides additional lift in the positive direction while performing its primary task of balancing the wing pitching moment. The aft-mounted (or conventional) horizontal tail must produce negative lift to do this job, as seen in Fig. 6-1. The wing, then, has to overcome the downward tail force as well as the weight of the airplane. With the horizontal surface mounted forward, as shown in Fig. 10-8, a nose-up balancing moment is provided by an upward lift. This lift adds to the wing lift and relieves it of some of its burden, thereby reducing drag. The end result is higher L/D of the overall airplane. The canard is also out of the wing downwash, and thus efficient at high angles of attack.

Fig. 10-7. Rutan VariEze canard.

At first this configuration appears to be very desirable; however, there are several disadvantages to the canard arrangement. The canard surface will increase in lift and, hence, in nose-up moment, with increased angle of attack: *destabilizing*. Stability can be provided by the wing, but this requires that the wing always be behind the CG in order to give a nose-down moment at increased angle of attack. With this arrangement, the wing cannot be allowed to stall before the canard or the airplane would pitch upward, deepening the stall, rather than tending to recover from it. On the other hand, if the canard stalls first, control is momentarily lost and a pitchdown motion will result. This dilemma requires a careful balance of lifting capability between the canard and the wing. Limiting the controls can be arranged so that the stall condition cannot be achieved under normal flight conditions. Unfortunately, this solution also robs the wing of some of its $C_L$ range, because there must be some margin below the actual maximum $C_L$ (stall point). This effect results in longer takeoff and landing distances.

Center of gravity travel is also rather limited with canard arrangements unless considerable area is designed into the canard for adequate control. Larger area detracts from stability; hence, canard area design becomes somewhat of a dilemma. Larger canard surfaces also contribute more induced drag, and the downwash reduces the wing effectiveness; thus, getting a net L/D reduction from this configuration is a complex design task.

With the horizontal surface mounted forward you might suppose that the aft portion of the fuselage could be shortened, and drag saved in this fashion. But, alas, the designer is thwarted again. The vertical tail must always be aft of the CG. An extended aft fuselage is usually required to support this surface and thereby achieve directional stability. Some designers have swept the wings and placed vertical fins at their tips. In this way the fin is located reasonably far aft of the CG. Rutan has used this arrangement rather successfully. Excessive sweep, however, reduces the low-speed effectiveness of the wing and again might defeat the intent of the basic concept in achieving high L/D.

Like everything else in the world of aerodynamics, the canard has some advantages and some disadvantages. It can be configured by careful design to yield a more efficient airplane. It is not without problems and certainly is not a panacea for the age-old quest for high lift and low drag. If it were, you would see a rapid shift to this type of configuration. (Contrary to popular belief, engineers are usually not hung up on traditional concepts.)

One other useful aspect of the canard arrangement has to do with engine location. It has long been known that the forward-mounted propeller is somewhat inefficient due to the interaction of part of the slipstream with the cowling. Pusher propellers do not have this problem, and are therefore more efficient. The pusher propeller requires an aft-mounted engine on a single-engine design and this shifts the CG pretty far aft. Because the wing must be near the CG, the wing must also be placed pretty far aft. The canard arrangement thus appears to be ideal for this type of configuration.

# FLYING WINGS

Another concept that is not new, but keeps popping up on the drawing boards, is the *flying wing*. Traditionally we house passengers and cargo in a fuselage. The fuselage produces no appreciable lift, but does contribute significantly to drag. It would seem logical, then, that if it could be eliminated, and everything contained in the wing, a very efficient design should result.

In theory this premise is true. Payload could even be distributed throughout the span of the wing so that the weight would subtract from the lifting force acting on the wing structure. The result would be much less support structure required and, consequently, less weight; however, the disadvantages again raise their ugly heads.

First of all, remember that airfoils have specific dimensions based on the chord length. In order to provide a maximum thickness of 6 feet using an airfoil of 15 percent thickness, the chord would have to be 40 feet. That is a pretty sizable wing and this points up a basic fact about flying wings: They are only applicable to very large aircraft. Otherwise, there is so little space available in the wing that a fuselage is necessitated.

Another major deterrent is the consideration of longitudinal stability and control. Airfoils can be made stable (given a positive pitching moment) if the trailing edge is bent upward, *reflexed*, as it is sometimes referred to. Such arrangement can eliminate the need for a horizontal tail. Another way of achieving stability in a flying wing is to sweep the wing and twist the tips to negative angles of attack, so that they act like a horizontal tail and fall sufficiently aft of the CG. Either of these methods can provide a degree of stability, although not great. Control is even harder to create with the short moment arms available. Consequently, flying wings cannot achieve high maximum lift coefficients, which makes for long takeoff and landing characteristics. Vertical tail location is also a problem, as with canards with short fuselages. The short moment arms available usually require extra-large surface area to provide adequate directional stability.

In summary, the flying wing again is one of those ideas that at first appears to have much merit, but on closer observation reveals itself to be wrought with problems. Some small "pseudo flying wings" have been somewhat successful. These are basically flying wings with a small fuselage, or pod, in the center to house the pilot. Very light aircraft—particularly sailplanes—that do not require high lift coefficients can be made controllable and occupiable in this fashion. On the other end of the scale, flying wings can be feasible for very large aircraft, if control problems can be surmounted. Special purpose aircraft, also, can sometimes benefit from the flying wing concept. The B-2 bomber for example takes advantage of the low radar cross section of this configuration. Designed as a stealth aircraft, meaning one that avoids detection, especially detection by enemy radar, the B-2 radically departs from previous bomber design concepts. The large flat surface is difficult to detect, can cause radar energy to be absorbed or be reflected in multiple directions, or present a very faint target on a radar screen, which are contrary to curved surface detection. The deep wing resulting from a long chord also allows for construction utilizing radar absorbent materials.

# EFFECTIVE ASPECT RATIO DEVICES

Throughout the discussion of aerodynamics and design, we frequently encounter a consideration of wing aspect ratio. Higher aspect ratio has the effect of reducing vortex-induced downwash and the resulting induced drag. Induced drag can be very significant under certain flight conditions, such as at low speed or high altitude. Range, climb, and takeoff and landing performance are all enhanced by high aspect ratio. It would seem very desirable to incorporate high aspect ratio into aircraft designs.

Unfortunately, high aspect ratio also has a drawback. Longer span places the air loads farther outboard, resulting in greater bending moments being placed on the wings. These higher bending moments must be resisted with heavier structure and the airplane ends up with higher weight. You reach a point in aspect ratio where the heavier weight creates more induced drag than the increased span can save.

It has therefore occurred to a number of enterprising designers of wings to create devices that would give the same result as increasing aspect ratio without physically doing so. This amounts to creating fake, or false, aspect ratio, but is more properly referred to as *effective aspect ratio*.

Because induced drag arises from the vortex flow around the wingtip, anything that can be done to restrict or impede this flow should reduce this type of drag. Earliest attempts at improvement in this area involved tapered or rounded tips (in the vertical view) to reduce the airloads at the tips. These efforts aided little because the airloads drop off rapidly toward the tip regardless of the planform shape.

## Tip shape effects

Early wingtips were often rounded, in keeping with the general aerodynamic rule of smooth transition from one surface to another. Developers soon recognized that interference of the flow around the wingtip was more conducive to low drag because the tip vortex causes induced drag. A sharp-edged tip (viewed from a frontal position) will tend to have this effect by causing a slight separation to the flow, somewhat analogous to a sharp leading edge causing early stall.

A more effective tip arrangement is the *drooped tip*. Drooped tips have been available for some time as add-on devices from various modification shops. They have also appeared on production airplanes, particularly some of the last Cessna singles that were produced, as shown in Fig. 10-9. Drooped tips give the spanwise flow a somewhat downward flow around the tip, as shown in Fig. 10-10. The result is a formation of the vortex farther outboard rather than at the actual wingtip. Because it is the distance between the tip vortices that determines the action of the vortices, the wing behaves in this respect as if it were wider in span; hence, the effective aspect ratio is increased without physically increasing it.

Computer analysis has been a boon to studies regarding the way that vorticity is shed across the wing and the way that it rolls up into two distinct tip vortices. Results of such research reveals that an even greater reduction of induced drag can be achieved by sweeping the tips upward, rather than downward as with the drooped tip. The theory is that the upward displacement of the vortex reduces its effect on the downwash, and, hence, the induced drag. This principle was employed by Cessna in twin-engine designs. Figure 10-11 shows a Cessna 310 wing with an upswept tip tank, which gives the tip flow greater upward motion.

The Hoerner wingtip is somewhat of a compromise in tips that reduce

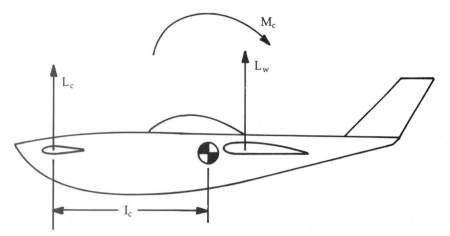

Fig. 10-8. Forces and moments acting about the center of gravity of a canard aircraft.

Fig. 10-9. Drooped wingtip on a modern Cessna.

induced drag, but retain simplicity of construction. This tip, designed by S. Hoerner, who did classic studies on aircraft drag, is used in a number of Piper airplanes, such as the Warrior tip shown in Fig. 10-12. It employs a slightly curved upsweep to the lower surface, which tends to boost the vortex both upward and outward, plus a sharp edge. It should be noted that none of these

Fig. 10-10. Spanwise flow around a drooped wingtip showing outward displacement of vortex center.

Fig. 10-11. Upswept tip tank on Cessna 310.

modified tip shapes produce a drastic reduction in induced drag, but in the aircraft design business, every little bit helps.

Very recent work by Germany's Richard Eppler has shown that the ideal shape of a wing, as viewed from the front or rear, that gives minimum drag, is that as shown in Fig. 10-13. The ideal angle of the tip section of the wing depends on the ratio of lift coefficient to aspect ratio, and is typically about 25 degrees. The anhedral angle of the section inboard of the tip section is much less—more on the order of 3 or 4 degrees. Bird watchers might recognize this shape as quite similar to that of the wings of eagles or swallows in gliding flight. It seems that nature had developed ideal aerodynamic designs long before we mortals could devise ways to compute them.

Fig. 10-12. Hoerner wingtip on Piper Warrior.

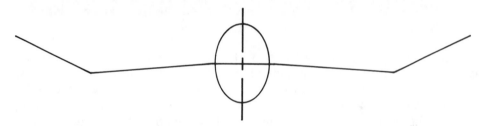
Fig. 10-13. Ideal dihedral configuration as proposed by Eppler.

## Endplates

Endplates were referred to in the discussion of ground effect in chapter 3. The endplate is a rather obvious device that should block the flow of air around the tip. It is simply a flat plate mounted against the wingtip. In order to be effective the endplate has to be quite large. In this case, it also creates parasite drag, mostly composed of skin friction drag. Unless the airplane is designed to operate almost entirely in a realm where induced drag predominates, the endplate ends up producing more total drag rather than less. They have been used to some degree of success on special-use aircraft such as low aspect ratio crop dusting airplanes.

An adaptation of the endplate principle is the tip-mounted fuel tank. If an external fuel tank is required due to reduced internal space, the tip seems to be the logical place to locate it. Because it is going to produce parasite drag no matter where it is placed, it might as well aid, in some degree, to reducing induced drag.

# Winglets

The most successful method of creating artificial aspect ratio to date is a recent NASA development known as a *winglet*. The winglet is a winglike shape mounted vertically on the tip of the wing, as shown on the latest model Learjet in Fig. 10-14. It appears as though the tip of the wing were actually bent to an upward position. It is, of course, a separate surface built into the wing structure.

Fig. 10-14. Gates Learjet Model 55 Longhorn showing winglets.

Another brainchild of NASA scientist Richard Whitcomb, the winglet is a clever device that takes advantage of the tip vortex flow to reduce drag. Normally, the vortex flow increases drag by downwash induction. Figure 10-15A shows the typical vortex flow around a wingtip. Figure 10-15B shows a top view of the wing with a winglet mounted on it. The inboard flow from the vortex combines with the normal free airstream flow to yield a resultant flow angled somewhat inboard, as shown.

The winglet is a small wing placed at an angle of attack with respect to this resultant flow. Lift is thus created on the winglet; however, lift is always perpendicular to the airstream and in this case the lift vector points somewhat in the forward direction. The forward component of this vector therefore gives a force in the direction of flight that is actually negative drag or thrust.

The winglet effect is not all gravy, however. The winglet creates parasite drag and because it is producing lift will have some induced drag associated with its own tip effect. There is also an interference effect at the juncture of the winglet and the wing. At low angles of attack (or low $C_L$ values) the tip vortex is relatively weak; hence, not much lift is produced by the winglet and it then contributes more drag than it saves. This situation is typical of cruise speeds at low altitudes.

Whenever vortex action is strong, the winglet becomes effective and reduces the drag more than it contributes to it. Such situations, which involve high lift coefficients, are encountered at low speeds or at very high altitudes. This explains why winglets have been first adopted by high-altitude designs such as the Lear 55 shown in Fig. 10-14. They are also useful on certain STOL aircraft where low-speed performance is of more concern than high speed.

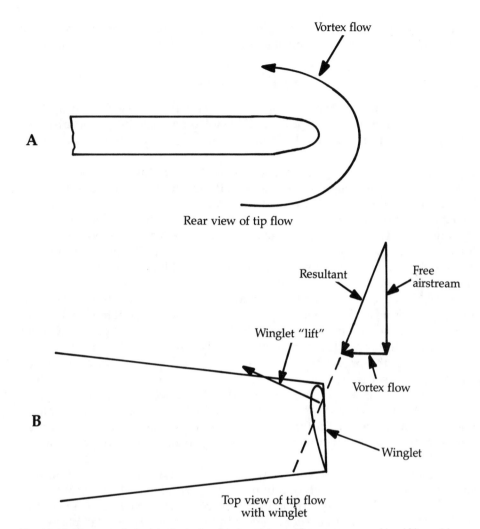

Fig. 10-15. Diagram of winglet effect, showing how inboard flow on upper surface (A) combines with free airstream to give resultant flow over winglet producing lift vector pointed somewhat forward (B).

One other important aspect of the winglet is the way in which it affects wing weight. The winglet does add to the wing bending moment and therefore requires a somewhat heavier wing; however, it has been shown that in order to achieve the same savings in induced drag by simply extending the wing straight out (resulting in higher actual aspect ratio), the wing weight would increase by three or four times over that with the winglet. It therefore appears to be a very efficient means of achieving increased effective aspect ratio.

Another application of the winglet has been on the tips of propeller blades. Propellers, remember, are really winglike surfaces producing lift in the forward direction. A winglet on the propeller tip has the very same effect as on the wingtip. Drag reduction on a propeller results in a higher propeller efficiency, which means that more of the engine horsepower gets converted into thrust.

# SWEPT WINGS

Throughout the book we have referred to various aspects of swept wings. Without repeating all of the details, it is desirable to summarize the characteristics of such wings because they are also a quite familiar sight on many modern airplanes.

The primary purpose of the swept wing is to reduce the local Mach number of the flow over the wing. This action enables the transonic airplane to cruise faster before encountering significant wave drag. This phenomenon is explained in chapter 7, and is the only really good reason for sweeping a wing excessively. Airplanes that fly approximately Mach 0.15 (100 knots) certainly do not require this treatment. Sweeping the wing actually reduces the lift coefficient for any given angle of attack. This effect is most noticeable at low speeds and results in longer takeoff and landing distances.

Sweep has been employed to advantage in low-speed aircraft for other purposes, when done to a moderate degree. One such use is to shift the overall aerodynamic center of the airplane to place it closer to the center of gravity without moving the entire wing. This approach is usually taken if the shift is found to be necessary after the prototype has flown, or if the airplane is made larger by extending the fuselage. The classic example of this type of fix is the famous DC-3 wing, which had its leading edge swept back in the DC-2 design for this very purpose. The aerodynamic center of the overall wing is really the aerodynamic center of the mean chord. So if the wing is swept back, the mean chord is shifted back, and its aerodynamic center (approximately its quarter chord point) moves back with it.

Another reason for sweeping the wing of a canard or a flying wing is to provide a point sufficiently far aft for mounting the vertical tail. The vertical tail has to be aft of the CG, and the farther aft it is placed, the smaller its area needs to be. With little or no fuselage extending behind the wing, vertical tail placement can be a problem. If the wing is swept enough, the tip can be far enough aft to provide a mounting point. Locating the vertical tail on the tip can improve the wing efficiency and it can also serve as a winglet to reduce induced drag. Many canard designs take advantage of this concept. The sweep of the wing also contributes to directional stability in itself and reduces the required vertical tail area, as discussed in chapter 6.

Currently there are a number of *forward-swept* wings appearing on the drawing boards. Forward sweep serves the same purpose as aft sweep for reducing the Mach number of the flow over the wing; however, forward sweep does not suffer the poor stall characteristics of the aft-swept wing. Aft-swept wings encounter spanwise flow that tends to force separation (and, hence, stall) at the tips first, due to the boundary layer build-up. Forward-swept wings do not have this problem because the spanwise flow is inboard and the wings prove to be much better in low-speed qualities.

Unfortunately, forward-swept wings have a serious structural problem. As the wing flexes upward under normal lift loads, the tip presents an exaggerated angle of attack to the airstream. This causes even greater lift, and because the center of lift is normally ahead of the center of flexure, it causes an extreme twisting moment on the wing. Under certain conditions, this twisting moment can cause structural failure. Until recently it was almost impossible to make the wing strong enough to resist such stress; however, with new composite structures, it appears to be quite feasible. Designers are attempting to take advantage of the good traits of forward sweep. Again, this configuration is of benefit only to airplanes in the Mach 0.8 or faster category.

# MODERN AIRFOIL DESIGN

The supercritical airfoils and their low-speed derivatives that Richard Whitcomb developed in the 1960s ushered in the serious application of computers to airfoil design. The low-speed airfoils of this era still assumed turbulent flow over most of the surfaces due to construction techniques of the time. While the GAW series of airfoils did achieve higher maximum lift coefficients than the 6-series, they did not improve on the drag at cruise angles of attack. In many cases they had more drag than these older airfoils. Many aircraft designers still opted for the old 4-digit NACA airfoils in lightplanes where maximum lift was the primary consideration because of the ease of construction. If high speed were the driving design parameter, the 6-series was relied upon due to the low drag characteristics.

In the 1970s, computer techniques became more refined, and new research was showing that laminar flow over a large portion of the airfoil was feasible, even at fairly high (but subsonic) speeds. Much of this work had been pioneered by Richard Eppler and F.X. Wortmann in Germany for application to sailplanes. NASA then began to take a serious look at developing an airfoil for general aviation that would combine the high lift of the early 4-digit airfoils and the low drag of the 6- series. The key was the achievement of true laminar flow over a significant portion of the airfoil strictly by the shape. No external blowing or other artificial means of preserving the laminar boundary layer was to be involved. These airfoils were termed *natural laminar flow* (NLF) airfoils. New construction techniques involving composites seemed to make this approach practical.

The old NACA airfoils were developed by altering one parameter at a time, such as thickness or camber, and then testing them in a wind tunnel. The resulting aerodynamic data was published in a catalog arrangement. The aircraft designer then had to pick one of these that most closely matched the design conditions for the proposed airplane. The new approach that NASA used was to first specify the desired characteristics that the design required, and then design an airfoil to meet these requirements. This approach is sometimes referred to as *inverse design*. Only high-capacity computers could make this technique possible.

NASA worked with the Eppler method, which began with a prescribed pressure (or velocity) distribution over the airfoil, and applied various aerodynamic theories to analyze the flow and to establish the boundary layer characteristics. A design condition is specified for an angle of attack, which depends, among other things, upon the weight, the forward speed, and the altitude (or density). One of the major advantages of the Eppler method is that it has provision to account for off-design conditions. The design condition is often cruise flight at best cruise altitude at full gross weight. If the design is optimized for this condition with no other considerations, the resulting airfoil might perform poorly in climb, or at other altitudes or weights. The method used by NASA was able to account for a variety of such conditions, and, thus, design an airfoil that is the best all-around airfoil for the intended aircraft.

The first such airfoil intended for light single-engine airplanes was the NASA NLF (1)-0416. The first two digits of the number (04) signify a design lift coefficient of 0.4, and the 16 refers to a maximum thickness of 16 percent of the chord. The 1 in parentheses refers to the first generation of such airfoils. A number of other NLF airfoils have been designed since then, including ones intended for specific aircraft. Figure 10-16 shows the shape of a typical NLF airfoil. This one, developed by NASA engineer Dan M. Somers, is designated the NLF (1)-0215F. The F stands for flapped because this particular airfoil was designed to accommodate flapping.

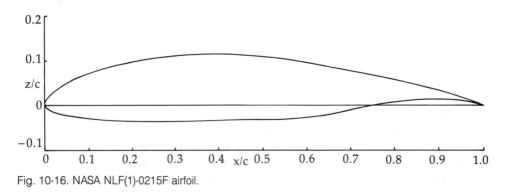

Fig. 10-16. NASA NLF(1)-0215F airfoil.

The natural laminar flow designation is restricted to airfoils that have laminar flow over at least 30 percent of the chord; however, with good fabrication methods, it is possible to achieve this condition over 50 to 70 percent. This phenomenon is achieved by a contour that continues to decrease the pressure over the surface toward the trailing edge for a considerable length of chord. This condition tends to draw the laminar flow aft, a *favorable pressure gradient*. High lift is attained by a substantially low pressure over a much greater length of chord than previous airfoils, which tend to have the low pressure concentrated in a "peak," or narrow area of chord, at the leading edge.

Natural laminar flow airfoils are beginning to show up on a number of new aircraft, such as the Piaggio Avanti (Fig. 10-17). The NLF airfoil used on this design is capable of maintaining laminar flow over approximately a third of the wing chord, and is a major part of the reason that this airplane is so efficient.

Fig. 10-17. Piaggio P180 Avanti.

# NEWER DESIGN DETAILS

A number of design details appear on recent aircraft that are somewhat different from conventional configurations. Swept vertical fins and T-tail arrangements are typical examples. These shapes are examined to some extent in chapter 8. While there is some aerodynamic advantage to the T-tail, it, like the swept fin, is primarily an attempt to make a more modern-appearing airplane. Such designs, like many of those of auto manufacturers, do little to improve performance. Airplane buyers, like consumers of automobiles and other devices, seem to crave something new and different in appearance. Designers of aircraft are severely limited in their flexibility in this area. Consequently, when a new style is possible that does not severely detract from performance, it usually becomes popular quite rapidly. Conservative designers usually stick with those configurations that prove to give best overall performance.

Some aerodynamic refinements are quite beneficial. The newer supercritical and NLF airfoils, for example, have many superior qualities to older airfoils. When properly applied to compatible designs they could give greatly improved performance. Winglets, improved cowl shapes, and carefully designed fillets fall into this category also.

# WHAT NEXT?

The highest priority for advanced aircraft development is in military designs; hence, most money and effort are put into this area. The newest concepts for military designs are kept secret for security purposes and cannot be discussed.

Aircraft for airline operation represent the next largest group in terms of research and development dollars. NASA and the aircraft manufacturers are considering the most feasible design concepts for the coming decades.

## Potential airliner designs

In recent years the driving parameter in new aircraft designs has been fuel efficiency. This is particularly true in the case of airline aircraft, where the fuel costs have risen to as high as 50 percent of the direct operating costs. Consequently, much effort has been directed at savings in this area. The Boeing 757 and 767 are good examples of aircraft designed with fuel efficiency greatly in mind.

Engine efficiency, of course, plays a major role; however, aerodynamic efficiency (read as lower drag) reduces the need for power, and can achieve the same result. Because airliners cruise at high altitudes where induced drag predominates, one area of concentration is in reduction of this offender. The logical approach is through high aspect ratio. NASA has proposed designs that would incorporate extremely high aspect ratio wings. To counteract the added structural weight of such designs, they have considered adding external struts (back to the 1930s!). At high altitudes, the parasite drag of the struts could be insignificant compared to the potential savings in induced drag. Several wings of fairly conventional span but thin chord have also been considered to increase aspect ratio. Both of these concepts could be combined with advanced winglet shapes to further combat induced drag.

Very narrow chord wings have also been considered for reducing skin friction drag. If a very smooth surface, narrow chord, and an airfoil with favorable pressure gradient characteristics could all be combined, a wing with nearly all laminar flow might be possible, even at very high speeds. Wings with nearly complete laminar flow could improve fuel efficiency of airliners by as much as 60 percent. A more possible means of achieving laminar flow is to reduce the boundary layer by suction over the wing surfaces. This approach, of course, requires some power and a rather complicated ducting and wing surface construction. Nevertheless, it has been proven quite effective in many actual flight tests.

Span loading is another concept that has potential in very large airplanes. This means distributing the payload over all, or nearly all, of the span. This technique reduces loads on the wing and results in lower wing structural weight. It is most efficient in a flying wing design. The flying wing is large enough to house payload—and thus becomes a very promising configuration—in airplanes of more than 1 million pounds gross weight.

Hypersonic transports are the rage at this writing. However, as explained in chapter 7, many problems have to be worked out before anyone will travel as a passenger at more than Mach 5. Whether this realm of flight becomes a reality remains highly speculative. Research efforts toward this end will be very extensive and very expensive. Nevertheless, the technology is progressing, and a New York to Tokyo flight in two hours remains a distinct possibility.

Futurists in airplane design have not ruled out supersonic transports, either. The Concorde is a very successful airplane, technologically. It suffers from economic problems due to its low payload capability, similar to those encountered in early transports way back in airline history; however, NASA has shown that with advanced technology available in coming decades, it will be possible to build large SSTs that are economically feasible. Such transports could bring in profit to the airlines without charging exceptional fares. These airplanes could do for supersonic transportation what the DC-3 originally did for air transportation and what the 707 did for subsonic jet transportation.

## General aviation

For years the lightplane pilot has been looking for the revolution in design that will provide an airplane that is both high in performance and affordable. Aircraft of this nature have often been promised—such as the $2,500 BD-1 of the early 1960s—but never delivered. Furthermore, it is highly unlikely that they ever will. In the current economic situation, aircraft manufacturers see profit only in the larger, expensive general aviation craft built for business use. Considerable development effort is put into this type of aircraft, however. There is the possibility that successful new technology will trickle down to smaller airplanes.

One of the most promising items of new technology for general aviation aircraft, as well as all aircraft, is the concept of *composite structures*. Composite simply means made up of different parts or materials. In the sense of aircraft structures, it refers to material made up of small, high-strength fibers embedded in a plastic-like medium that bonds and shapes them. Fiberglass-reinforced plastic is really a composite material. The fiberglass strands actually provide the strength, while the plastic resin holds them in place and allows the material to be formed into the desired shape.

Fiberglass is rather heavy, however. The newer composite materials used in aircraft structures employ boron or carbon fibers. Such materials provide very high strength with very low weight, the goal of structural engineers since the dawn of flight. Weight savings of 20 to 30 percent over conventional aluminum alloy structures are quite feasible with composites.

The other advantage of composite materials is that it can be molded and bonded to provide a very smooth aerodynamic surface. Plywood skins were used for many years simply because they could be bonded and thus free the surface of drag-producing rivet heads. Composites can be made even smoother and much stronger. Such materials, when used with airfoils with favorable pressure characteristics, have the potential for providing natural laminar flow at fairly high speeds. NASA has already made considerable progress with research in this area, as we noted.

Major breakthroughs in performance, of course, follow revolutionary developments in propulsion systems. Lightplane fans have also been anticipating news in this area for a long time. On numerous occasions, engine manufacturers have set up special task forces to investigate the possibility of a low-cost light turbine engine, or jet. Such efforts have always failed. Turbine engines require exotic materials and complex construction techniques to build turbines that hold up under the extreme temperatures that they must encounter. Recently there has been some research with ceramic blades that could reduce engine costs some; however, even this technique is not sufficient to provide small engines at anywhere near the cost of equivalent piston engines.

One type of powerplant that does show some promise for lightplane application is the *rotary engine*. This engine, often called a *Wankel* after its inventor, Felix Wankel, is an internal combustion engine but without pistons. The combustion process drives the rotor in a continuous, smooth, rotating manner. The reciprocating motion of pistons wastes considerable energy, particularly when you consider that only one of four strokes actually produces power for thrust.

Wankels have been used in automobiles with somewhat limited success, hindered especially by troubles experienced with sealing the rotor against the chamber walls. This process was particularly troublesome at low engine speeds, where cars frequently operate. The problem was much less severe at high speeds where aircraft engines normally operate. Also, the first rotaries were put into cars before their development was very far along. Much refinement has been done to the design in recent years.

Another big advantage of the Wankel is that it can run on a variety of fuels, much like jet engines. This might be a very important feature in future decades if petroleum fuel supplies become exhausted. Alcohol and other synthetic fuels have already performed quite satisfactorily in these engines.

Unfortunately, the Wankel does not produce a tremendous amount of power for its weight. While it might prove to be better than piston engines, and more fuel-versatile, it will not be to light airplanes what the jet was to military and transport aircraft.

Turboprop powerplants will undoubtedly be the prime mover of high performance business and commuter aircraft for quite some time. With the importance of fuel economy in the modern world coupled with the reliability of current turbine engines, turboprops are beginning to appear on single-engine designs. The Beech Lightning, the Smith Prop-Jet, and the Interceptor 400 were apparently premature attempts at such development. Piper attempted to put a turboprop in the Malibu, and OMAC came close to completing the Laser 300, but at this writing that company has reorganized, and is struggling to get back into active development.

Fortunately, the Cessna Caravan utility airplane has been very successful with a single turboprop, and recently the French introduced the Aerospatiale TBM 700 (Fig. 10-18). This airplane is an all-new design, highly refined by computer-aided design, and capable of 300 knots cruise, more than 1,000 nm range,

Fig. 10-18. Aerospatiale TBM700.

and is certified for flight to 30,000 feet; however, with a price tag of more than $1 million, such aircraft are not for the weekend pilot.

Other attempts at revolutionary designs for the business traveler have been in the works over the past decade or so. While not a single-engine, but a single-propeller-driven airplane, the Learfan promised a refreshingly new approach to efficient aircraft (Fig. 10-19). This airplane was the last idea conceived by the fabulous Bill Lear (of Learjet fame) before his death. It was carried on to prototype by his widow and flew on January 1, 1981. Unfortunately, the development of a

Fig. 10-19. The Lear Fan 2100.

revolutionary aircraft and a revolutionary construction method was too over-whelming for a small company to handle. The project terminated when funding sources were totally exhausted. The airplane was made entirely of graphite-epoxy composite, and was pioneering the certification process for composite air-craft.

Meanwhile, Beech Aircraft developed the radical, new, twin-engine canard known as the Starship (Fig. 10-20). This aircraft is also of totally composite con-struction and continued the pioneering effort to become the first such aircraft to be certified. It was designed with help from Burt Rutan, famous for his home-built canard designs. Some of the chronic stability and control problems of the canard configuration have been overcome in this design by a variable sweep-angle of the canard surface.

Fig. 10-20. Beech Starship.

Piaggio of Italy developed the three-lifting-surface design known as the Avanti (Fig. 10-17). This aircraft (originally to have been a joint effort with Gates Learjet) combines some of the advantages of the canard and conventional tail arrangement. Tests have shown that the three-surface arrangement produces less induced drag than a pure canard design, while retaining the positive lift of the canard to overcome wing pitching moment. This configuration, aided by exces-sive use of computer-aided design methods, has resulted in a very low-drag air-plane. The fuselage has extensive laminar flow, as well as the wing, as a result of the use of such technology. The Avanti is also a hybrid in construction. Primarily made of aluminum alloy, it also has a significant amount of composite parts, particularly in the tail and forward wing. This approach has allowed for a light empty weight, which also reduces induced drag.

So, it would seem that the long-hoped-for revolution in airplane design might still come to pass; however, all of these airplanes are definitely in the luxury class and offer little solace to the sport pilot. If they succeed, the technology will eventually filter down to smaller aircraft, but it is highly unlikely that there will be any great breakthroughs in cost reduction.

For a while, the ultralight appeared to be the greatest hope for the sport pilot. Unfortunately, many of the first such airplanes had extremely high drag, questionable handling qualities, and unproven structural integrity. Later designs improved upon many of these faults, but, in the meantime, the ultralight had gotten a bad name from the very high accident record, and sales dropped rapidly. Even the good designs are very limited in utility, with low speed, short range, and intolerance of even moderate winds. They remain a very specialized class of airplanes.

Major manufacturers have all but given up on single-engine piston production. At this writing, only Piper is turning out a significant number of such aircraft, and their continued success in this area is very questionable. The greatest promise for innovation in sport aircraft seems to lie in the design of kits for homebuilt aircraft. At one time, homebuilts consisted largely of biplanes and other fun-type aircraft that were mostly throwbacks to an earlier, more romantic era of flight. No more. Some of the most advanced light airplanes are coming from the drawing boards (or, more correctly, computers) of the kit designers. The composite Lancair and Glasair are good examples of such aircraft. The Glasair is available in retractable or fixed gear (Fig. 10-21), and is capable of cruising up to 200 knots with 180 to 200 horsepower. The Glasair III, a stretched version with 300 horsepower, will fly faster than 230 knots.

Even more speed has been achieved with the conventional all-metal Questair Venture. Designed by a former Piper engineer, it incorporates a scaled-down copy of the Malibu wing and tail. With a large 270-horsepower engine, this airplane can achieve more than 240 knots at cruise, and climb more than 2,000 fpm. Its low drag is achieved partially by a very short fuselage, which reduces wetted area, and, hence, skin-friction drag. Stability is maintained by a rather large tail. Computer technology has enabled the establishment of this configuration as the best compromise for overall greatest efficiency.

A unique approach to modern airplane design is to take an older aircraft and completely redesign it. This is what Roy LoPresti did in creating the SwiftFury (Fig. 10-22). He chose the old Globe Swift, which was a sleek looking airplane, but really was not so "swift" for its power. LoPresti's philosophy on this design was to get the most power possible out of a relatively small engine, and to optimize the shape for lowest drag, achieving speed by a two-frontal attack.

Useful horsepower is increased by an alternate ram-air induction channel, which bypasses the filter, and is selectable by the pilot. This air is further energized by indexing the propeller to provide a pulse of air from a passing blade just as the intake valve for a cylinder opens. Drag cleanup was enhanced by a com-

Fig. 10-21. Glaisair kitplane.

Fig. 10-22. LoPresti Piper SwiftFury.

puter technique to optimize the shape of such things as the cowl, windshield, and wing fillets. Gear doors were designed to completely enclose the wheel and brake assembly, as well as close after extension to improve the wing's performance at low speeds. The result was a vastly improved machine, which flies much faster than 200 mph on 200 horsepower. It is designed as a production airplane, but whether that will come to pass or not remains to be seen.

Revolution in airplane design is rare, as we have seen. New technology does evolve, however, and this, along with the better application of old technology, will continue to give us better airplanes in future years. Despite the rising costs of almost everything, mankind will somehow find a way to keep the dream of flight a continuing reality.

# Appendix A

# Aerodynamic data
# for four
# NACA four-digit airfoils
# and
# Properties of the
# Atmosphere

NACA 1412 Airfoil Lift Characteristics

Fig. A-1. Aerodynamic characteristics of the NACA 1412 airfoil section.

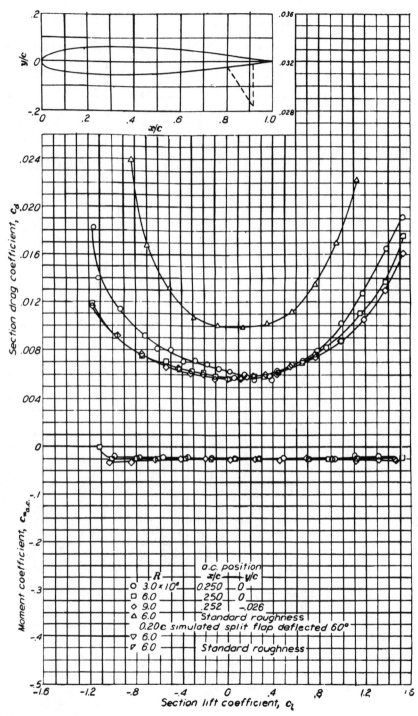

NACA 1412 Airfoil Drag and Moment Characteristics

Fig. A-1. Continued.

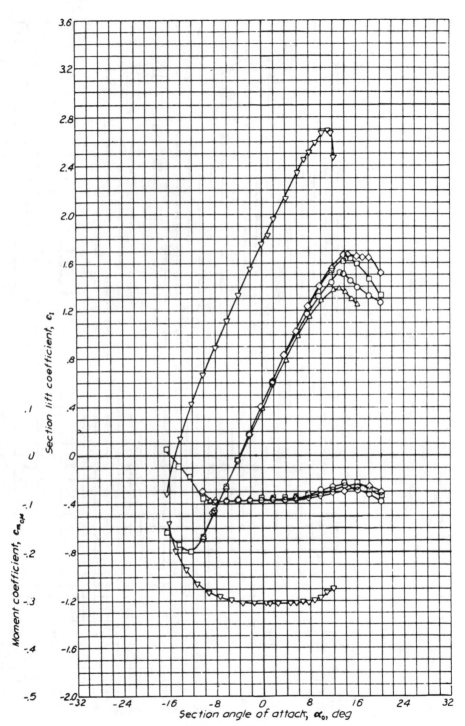

Fig. A-2. Aerodynamic characteristics of the NACA 4412 airfoil section.

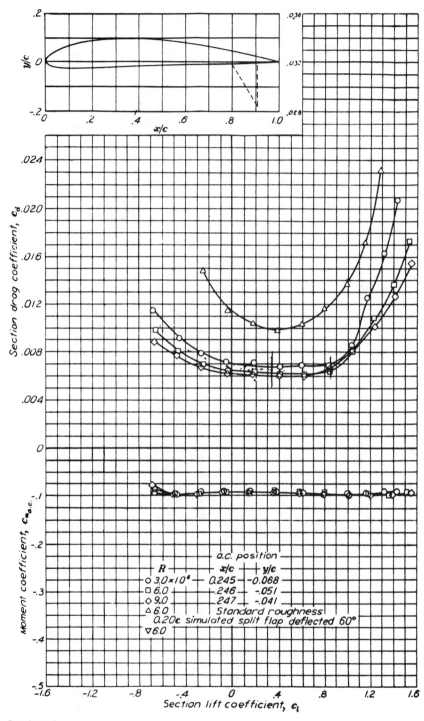

NACA 4412 Airfoil Drag and Moment Characteristics

| | R | x/c | y/c |
|---|---|---|---|
| ○ | 3.0×10⁶ | 0.245 | -0.068 |
| □ | 6.0 | .246 | -.051 |
| ◇ | 9.0 | .247 | -.041 |
| △ | 6.0 | Standard roughness | |

a.c. position

0.20c simulated split flap deflected 60°
▽ 6.0

Fig. A-2. Continued.

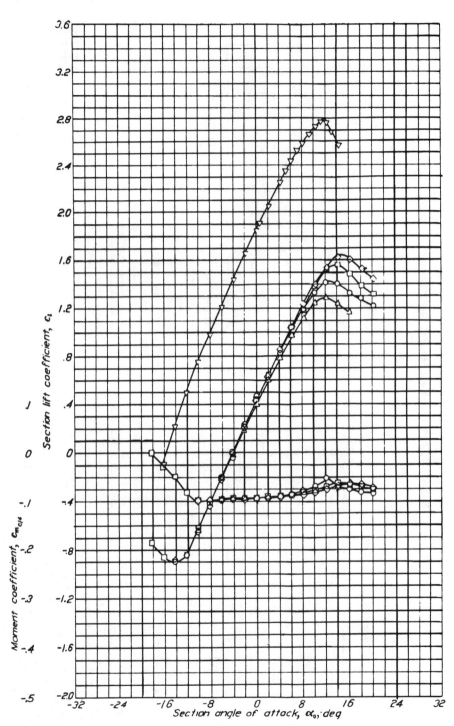

Fig. A-3. Aerodynamic characteristics of the NACA 4415 airfoil section.

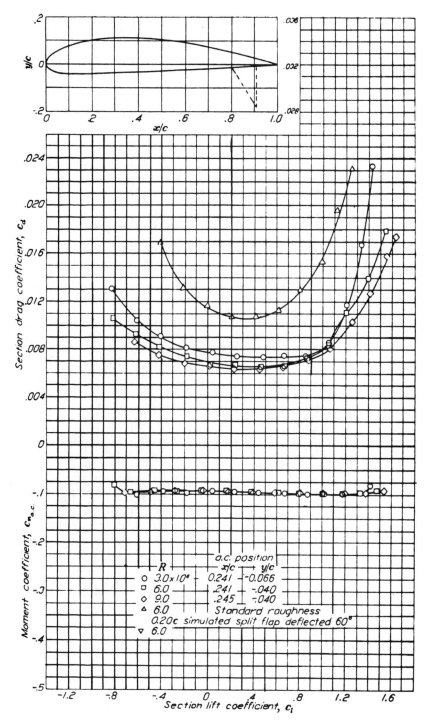

Fig. A-3. Continued.

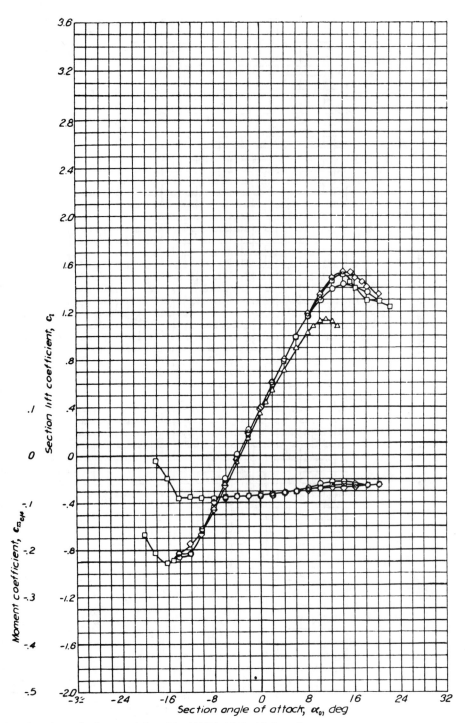

Fig. A-4. Aerodynamic characteristics of the NACA 4418 airfoil section.

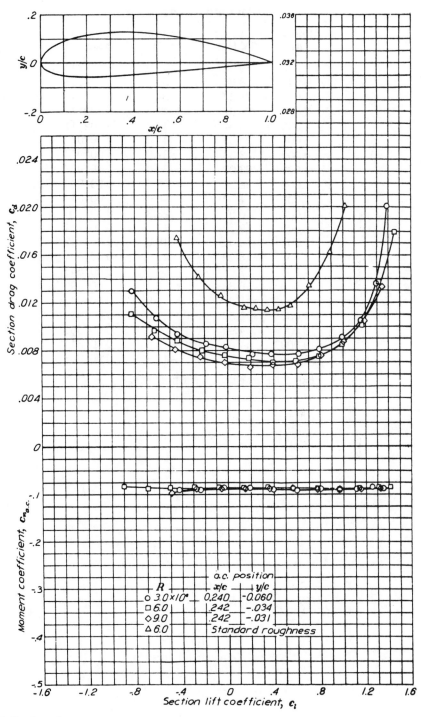

NACA 4418 Airfoil Drag and Moment Characteristics

| R | a.c. position | |
|---|---|---|
| | $x/c$ | $y/c$ |
| ○ $3.0 \times 10^6$ | 0.240 | -0.060 |
| □ 6.0 | .242 | -.034 |
| ◇ 9.0 | .242 | -.031 |
| △ 6.0 | Standard roughness | |

Fig. A-4. Continued.

## Standard Atmosphere

| Altitude (ft) | Pressure (in. Hg) | Temperature (°F) | Density (slug/ft$^3$) |
|---|---|---|---|
| 0 | 29.92 | 59.0 | 0.002377 |
| 1,000 | 28.86 | 55.4 | 0.002308 |
| 2,000 | 27.82 | 51.9 | 0.002241 |
| 3,000 | 26.82 | 48.3 | 0.002175 |
| 4,000 | 25.84 | 44.7 | 0.002111 |
| 5,000 | 24.89 | 41.2 | 0.002048 |
| 6,000 | 23.98 | 37.6 | 0.001987 |
| 7,000 | 23.09 | 34.0 | 0.001927 |
| 8,000 | 22.22 | 30.5 | 0.001869 |
| 9,000 | 21.38 | 26.9 | 0.001811 |
| 10,000 | 20.57 | 23.3 | 0.001756 |
| 11,000 | 19.79 | 19.8 | 0.001701 |
| 12,000 | 19.02 | 16.2 | 0.001648 |
| 13,000 | 18.29 | 12.6 | 0.001596 |
| 14,000 | 17.57 | 9.1 | 0.001545 |
| 15,000 | 16.88 | 5.5 | 0.001496 |
| 16,000 | 16.21 | 1.9 | 0.001448 |
| 17,000 | 15.56 | − 1.6 | 0.001401 |
| 18,000 | 14.94 | − 5.2 | 0.001355 |
| 19,000 | 14.33 | − 8.8 | 0.001311 |
| 20,000 | 13.74 | −12.3 | 0.001267 |

Fig. A-5. Properties of the atmosphere.

# Appendix B

# Review questions, problems, and answers

## QUESTIONS AND PROBLEMS

### Chapter 1

1. A balloon rises from the earth's surface and ascends to 10,000 feet altitude, where it remains. Why does it do this?

2. How did Cayley's approach to achieving manned flight differ from da Vinci's approach?

3. What was the key to the Wright brothers' success?

4. What did scientists such as Prandtl, Joukowsky, and Lanchester contribute to the development of the first airplane?

### Chapter 2

1. The airflow in a venturi tube has a velocity of 100 ft/sec in the wider part where the cross-sectional area is 6 sq ft, and the tube narrows down to an area of 4 sq ft. The air density is 0.002378 slugs/sq ft. What is the velocity at the 4 sq ft section, and what is the dynamic pressure at this point?

2. Why does a cambered airfoil produce a pitching moment when it produces lift, and in what direction is the moment?

3. What is the maximum camber and the maximum thickness of a 60-inch chord wing using a NACA 4415 airfoil?

4. What lift coefficient would give the lowest drag for a $63_2$-418 airfoil?

5. What is the area of a wing that has an aspect ratio of 8 and a span of 30 feet?

6. An airfoil produces a lift coefficient of 0.6 at 4° angle of attack, but when used in a wing with an aspect ratio of 8, requires 5.5° to produce the same lift coefficient. Why, and what is the additional 1.5°?

7. How much lift would the wing shown in Fig. 2-43 produce at 10° angle of attack in a dynamic pressure of 25 lb/sq ft and a wing area of 100 sq ft?

8. The wing in Fig. 2-43 stalls at about 14.5° angle of attack. Where would you expect it to stall with flaps deflected; the same angle, greater, or lesser angle?

## Chapter 3

1. What are the types of boundary layers, and which produces the most drag?

2. Why do golf balls have dimples?

3. An antenna of 0.05 sq ft of cross-sectional area produces 1 pound of drag in an airflow with a dynamic pressure of 25 lbs/sq ft. What is its drag coefficient?

4. An airplane's gross weight is increased by 20 percent; how much more induced drag will it produce in steady level flight?

5. What could be done to redesign the above airplane so that the induced drag would not change with the 20 percent weight increase?

6. An airplane produces 300 lbs. of drag in an airflow with a dynamic pressure of 30 lb/sq ft. What is its equivalent flat plate, or parasite, area?

7. What is the parasite drag coefficient of the above airplane if the wing area is 200 sq ft?

8. What are the different types of drag that make up parasite drag?

9. When an airplane is flying at a height of 20 percent of its wingspan, what is the reduction in induced drag due to ground effect over that at a significantly higher altitude?

## Chapter 4

1. Why is the propeller a more efficient producer of thrust than the jet engine at low speeds?

2. What is a turbofan engine?

3. What can be done to an existing engine to cause it to produce more power?

4. What can be done in the redesign of an engine to make it produce more power?

5. How does a supercharger increase power output at altitude?

6. If we reduced the rpm of an engine, but maintained the same forward speed (such as by diving), how would the propeller pitch have to change to keep it at its most efficient angle?

7. What is the most efficient type of aircraft powerplant at a cruising speed of about 500 mph?

## Chapter 5

1. Figure B-1 depicts the power curves for a fictitious airplane, including the power required curve, and the power available curves for 100 percent power and 75 percent, or cruise, power. From this chart, determine the following:

   A. Maximum level flight speed
   B. Cruise speed at 75 percent power
   C. Maximum range speed
   D. Maximum endurance speed
   E. Maximum glide distance speed

2. If the airplane in question 1 weighs 2,000 lb, what is its rate of climb at 70 knots?

3. If the above airplane were a twin, and lost one engine, what would be its rate of climb at 70 knots?

4. What is the maximum specific range of the airplane in question 1, if the SFC is 0.5 lb/bhp-hr and the propeller efficiency is 75 percent?

5. An airplane has a maximum L/D ratio of 10. How far can it glide from an altitude of 6,000 ft?

6. An airplane breaks ground on takeoff at 80 knots in 1,200 ft in no-wind conditions. How long is the takeoff with a 20 knot headwind?

7. What is the pilot's emergency procedure if one engine of a twin fails before $V_1$ is reached? What if it fails after $V_1$?

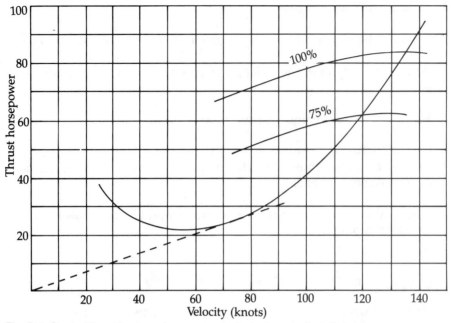

Fig. B-1. Chart of thrust horsepower versus velocity for hypothetical airplane.

8. Approximately how much longer is the braked portion of the landing roll if the runway is encountering heavy rain as opposed to a dry condition?

9. How would maneuvering speed change if gross weight were reduced by 16 percent?

10. An airplane has a speed of 480 knots at sea level. Without any more power applied, how high could it go, if zoomed to 200 knots?

## Chapter 6

1. In what direction is the pitching moment for a cambered airfoil, and in what direction is the tail trim load to balance it?

2. If an airplane is statically stable longitudinally, what sort of dynamic stability is possible?

3. Why does trim tab deflection cause the primary control surface to move in the opposite direction?

4. How does dihedral provide sideslip stability?

5. What causes adverse yaw?

6. Why does rudder deflection only cause a rolling moment?

7. What can be done to a design to reduce Dutch roll tendency, and what other effect does this modification have?

8. At what conditions would an airplane make the highest rate of turn, and at what conditions would it make the smallest radius of turn?

9. What load factor, or g-force, does an airplane pull at 40° angle of bank?

10. If the airplane in the above question stalls at 60 kts in a 1 g condition, what would be its stall speed in the 40° bank?

## Chapter 7

1. An airplane is going 610 mph at an altitude where the temperature is 59°F. What is its Mach number?

2. What is critical Mach number?

3. What would be the chordwise Mach number hitting a wing that is swept 25°, if the airplane is moving at Mach 0.8 with respect to the free airstream?

4. How does the angle of the shock wave that a body creates in supersonic flow relate to its Mach number?

5. What is the function of vortex generators located ahead of a control surface?

6. What is the basic principle of the area rule?

7. Why is Mach 5 considered to be the boundary of hypersonic flight?

8. What is the most promising engine for hypersonic flight?

## Chapter 8

1. Perform the preliminary sizing of a fixed-gear, low-wing airplane to meet the following specifications:

   • Payload: 2 persons (including pilot), plus 75 pounds of baggage
   • Cruise: 130 kts at 7,000 ft
   • Range: 600 nm

   Determine the gross weight, wing area, wingspan, engine horsepower, and fuel capacity.

2. Determine the horizontal tail area for the above airplane to give about the same stability as the Piper Arrow, if the tail moment arm is determined to be 12 ft.

3. Determine the vertical tail area for the same airplane as above considering the same tail moment arm, and an average tail volume coefficient.

## Chapter 9

1. Why is the fan located downstream of the test section in an open circuit wind tunnel?

2. A wing model of 0.5 ft in chord and 4 ft in span is tested in a wind tunnel at sea level with a velocity of 80 ft/sec. If the lift measures 7.6 lb, what is the lift coefficient?

3. What is the Reynold's number of the above wing as tested?

4. What three basic types of tests are performed in a wind tunnel, and what is a common method of measuring each?

5. How can shock waves be made visible in a supersonic tunnel?

6. Suppose that you have designed a wheel fairing for your homebuilt. How could you flight test it to see if it is efficient in reducing drag?

## ANSWERS
## Chapter 1

1. The gas inside the balloon is less dense than the surrounding air at sea level; hence, there is an imbalance of pressure forces due to the weight of the balloon plus the pressure on the top of it not being equal to the pressure on the bottom. At 10,000 feet, the air is less dense, or lighter for a given volume, so that, at this altitude the balloon is the same weight as the amount of air that it displaces. All forces are in balance and there is no longer a net force to move the balloon upward.

2. Cayley separated the functions of producing lift and thrust, concentrating on a fixed wing moving through the air to produce lift. Da Vinci thought that man could achieve flight by flapping wings to produce both lift and thrust simultaneously, as birds do.

3. The Wright brothers used a systematic, scientific approach to study all aspects of flight. They addressed each of the following areas in depth prior to attempting powered flight:

- Aerodynamics, emphasizing airfoils with best lift-to-drag ratio.
- Development of a light-weight engine with sufficient power.
- Development of an efficient propeller.
- Understanding stability and control to develop adequate controls.
- Teaching themselves to handle the controls.

It was the last two areas in particular that differed from many previous attempts at flight. Many early experimenters thought that the production of adequate lift and thrust was all that was necessary, and that the airplane would essentially fly itself, once airborne.

4. Aerodynamic scientists developed the theory of winged flight about a decade after the first airplane flew. It was well into the 1920s that theory began to be used in designing aircraft. The first developments relied almost entirely on experimentation.

## Chapter 2

1. From the continuity law, the area times the velocity at each point must be constant. In the wide section, 6 times 100 equals 600 cu ft/sec; thus in the narrow section, the velocity times 4 sq ft must also equal 600, and could be found by dividing 4 into 600 to yield 150 ft/sec.

$$V = 600/4 = 150 \text{ ft/sec}$$

Dynamic pressure is $1/2$ times the density times the velocity squared, or,

$$q = 1/2 \times 0.002378 \times 150^2 = 26.75 \text{ lb/sq ft}$$

2. When a cambered airfoil produces lift, the center of pressure on the top surface is always farther aft than the center of pressure on the bottom surface. The resulting pressure forces produce a twisting moment tending to move the leading edge (or nose) down.

3. The first digit indicates that the maximum camber is 4 percent (or 0.04) of the chord. This fraction times the 60-inch chord would yield 2.4 inches. The last two digits indicate a maximum thickness of 15 percent, or 0.15 times the chord, which is 9 inches.

4. The 4 in this 6-series airfoil designation indicates that the optimum lift coefficient (that for least drag) is 0.4.

5. Aspect ratio is defined as span squared divided by area; therefore, the area would be the span squared divided by the aspect ratio, or, 30 squared divided by 8, which is 112.5 sq ft.

6. The wing with some actual span produces downwash, which reduces the amount of lift at a given angle of attack, and has the effect of reducing the slope of the lift curve, as shown in Fig. 2-36, The additional 1.5° needed to achieve the same lift coefficient as the airfoil section is referred to as the "induced angle of attack".

7. From Fig. 2-43, we can see that the wing produces a lift coefficient of about 1 at 10° angle of attack. Lift is defined as lift coefficient times dynamic pressure times wing area; thus, 1 times 25 times 100 is 2,500 lb.

8. Trailing edge flaps increase the maximum lift coefficient of a wing, the point where the wing stalls, but this occurs at a lesser angle of attack than the unflapped wing, as shown in Fig. 2-48.

## Chapter 3

1. Boundary layers are either laminar or turbulent, depending on the speed, viscosity, and length of the flow, and the surface roughness. Because the turbulent boundary layer is thicker, it produces more drag.

2. Even though turbulent boundary layers have more skin friction drag, they have more energy, which can delay separation at low Reynolds numbers, and thereby reduce the pressure drag. The net effect of both pressure and skin friction drag could be lower with turbulent flow than with laminar. This is the case with golf balls, and the dimples serve to induce the turbulent flow.

3. The drag coefficient is the drag divided by the dynamic pressure and the area. Thus, 1 divided by 25 and by 0.05 is 0.08.

4. If the weight is increased by 20 percent, the new weight is 1.2 times the old weight. Because induced drag is proportional to the square of the weight, the new induced drag would be 1.2 squared, or 1.44, times the old induced drag, or 44 percent more.

5. Because induced drag is also inversely proportional to the square of wing-span, an increase of 20 percent in wingspan would cancel out the increased weight.

6. Equivalent flat plate area is the drag divided by dynamic pressure. Thus, 300 divided by 30 would yield 10 sq ft.

7. Parasite drag coefficient of an airplane is usually based on the wing area, and is the equivalent flat plate area divided by the wing area. Hence, 10 divided by 200 is 0.05.

8. Primarily, parasite drag is made up of pressure drag and skin friction drag. However, special drag types may be included also, such as cooling, interference, etc. Generally, it is all drag except that due to lift.

9. From Fig. 3-20, it can be determined that at 0.20 of wingspan, the induced drag is 0.70 times the normal induced drag. Thus, it is reduced 0.30 or 30 percent.

# Chapter 4

1. Thrust is produced by accelerating a mass of air; however, kinetic energy is lost in the accelerating process, proportional to the degree of acceleration. The most efficient way of producing thrust is to use a large mass of air and accelerate it only a small amount. The propeller has the capability of providing acceleration to a larger mass than the jet, and, it thus requires less velocity increase.

2. A turbofan is a turbojet with a ducted fan mounted on the shaft, which accelerates a mass of air that passes outside of the burner section. This air does not get the high acceleration of that which passes through the burners and, thus is accelerated more efficiently. High thrust does come from the gas emitted from the hot section, and thus, the turbofan is somewhat of a compromise between a jet and a propeller.

3. Because power comes from pressure, length of stroke, area of cylinder, and rotational speed (PLAN), the simplest things are to increase rpm or the air pressure in the intake manifold.

4. If the engine were to be redesigned for more power, the designer could increase the area of the cylinders or the length of the piston stroke, giving higher volume. Also, the air pressure in the cylinder could be increased by increased compression ratio.

5. The supercharger compresses the air in the intake manifold, giving air at altitude more pressure than is available in ambient conditions.

6. From Fig. 4-29, we can see that reduced rpm with constant forward speed would decrease the angle of attack; therefore, the pitch would have to increase to maintain the original angle of attack.

7. From Fig. 4-33, we see that at 500 mph, the turbofan has the highest efficiency. At this speed, the propeller efficiency has dropped below the turbofan, and the pure jet has not yet reached it.

# Chapter 5

1. A. The power required and maximum power available curves cross at 135 knots; hence, this figure is the maximum level flight speed.
   B. The 75 percent power available curve crosses the power required at 120 knots, making this the cruise speed.
   C. A line from zero to the tangency point of the power required curve just touches at about 70 knots, which determines the maximum L/D speed, the speed for maximum range.
   D. Maximum endurance occurs at the minimum power required, which is about 55 knots.
   E. Maximum glide distance also occurs at the maximum L/D speed of 70 knots.

2. The maximum power available at 70 knots is 68 hp, and the power required at this speed is 24 hp. An excess power of 44 hp, thus, exists, which, when divided by the weight of 2,000 lb and multiplied by the conversion factor of 33,000, yields a rate of climb of 726 ft/min.

3. If one engine were lost, the power available would be cut in half, or reduced to 34; however, the power required would still be 24; thus, rate of climb would be 34-24 divided by 2,000 and multiplied by 33,000, or only 165 ft/min.

4. Specific range is velocity divided by SFC times BHP. It was already determined in question 1 that the maximum range would occur at 70 kts. The BHP is THP divided by propeller efficiency, or 24 hp/0.75, which equals 32. The solution then is 70 divided by 0.5 times 32, which is 4.375 miles per pound of fuel.

5. The L/D ratio is also the glide ratio. Thus, 10 times 6,000 feet would give a glide distance of 60,000 ft. or approximately 9.87 nautical miles.

6. The no-wind takeoff distance is multiplied by the correction factor given as the square of $(V_o-V_w/V_o)$. Subtracting 20 from 80, dividing by 80, and squaring the result, yields a correction factor of 0.5625, which, when multiplied by 1,200, gives a new takeoff distance of 675 ft.

7. If an engine fails before $V_1$, the pilot should have sufficient runway available to abort, so power is cut and brakes applied immediately. If an engine fails after this point, there is not sufficient runway available to stop, so full power is applied to the remaining engine (if not already on) and takeoff is continued.

8. The friction coefficient is reduced from 0.7 to 0.3, a reduction of 0.4, or 40

percent. A reduction in friction coefficient by this much would increase the landing roll by about the same amount.

9. Maneuvering speed varies as the square root of the gross weight; thus, a reduction in weight of 16 percent would reduce maneuvering speed by 4 percent.

10. From Fig. 5-29, the airplane would have an energy height of about 10,000 feet; thus, at 200 kts the airplane would be able to reach a little over 8,000 ft (actually calculated to be 8,226 ft).

## Chapter 6

1. Because the center of pressure of the upper surface is farther aft than that of the lower surface for a cambered airfoil, the pitching moment is nose down; therefore, the tail load must be down to create a nose up pitching moment to balance the airplane.

2. If the airplane is statically stable it could be dynamically stable, unstable, or neutrally stable.

3. The trim tab is located well aft of the hinge line of the primary surface. Therefore, even though the trim tab is small, a force created by its deflection has a relatively long moment arm, which creates sufficient moment to rotate the primary surface about its hinge line. A downward movement of the tab, for example, would create an upward lifting force, and the resulting moment would rotate the primary surface upward.

4. If an airplane with dihedral rolls and slips in the direction of the downward wing, the relative wind against that wing would have a vertical component, which would tend to roll the airplane in the opposite direction.

5. In initiating a bank, the downward-deflected aileron on the wing to be raised creates more drag than the upward-deflected aileron on the wing to be lowered. As the turn progresses, the wing into the turn produces less drag because it is moving slower than the opposite wing. Both situations cause a yawing moment opposite to the direction of bank, the intended turn direction.

6. Rudder deflection initially causes a yawing moment, which means that the wing away from the direction of yaw will move faster than the one into the yaw. Faster speed on the outward wing means more lift, and less speed on the wing into the yaw means less lift; hence, a rolling moment is created.

7. A larger vertical tail will reduce Dutch roll tendency; however, this will also increase its tendency for spiral instability.

8. The steepest permissible bank angle (60° if it is not acrobatic) and the slowest possible speed (limited by stall) will give the highest rate of turn. These same conditions yield the smallest turning radius.

9. The load factor is 1 divided by the cosine of the bank angle. This calculation would yield about 1.3 for 40°. This result can also be seen graphically on Fig. 6-23.

10. The stall speed increases by the square root of the load factor. The square root of 1.3 is about 1.14, meaning that the stall speed increases by 14 percent. This result is also illustrated in Fig. 6-24. Multiplying 60 kts by 1.14 would give about 68 kts for this particular case.

## Chapter 7

1. As noted in the first section of this chapter, 59°F is the standard temperature at sea level, and the speed of sound there is about 761 mph. Mach number is true airspeed divided by the speed of sound, or 610 divided by 761, in this case, which is 0.80.

2. Critical Mach number is the Mach number of the free airstream when the speed of the airflow first reaches Mach 1 somewhere on the airframe surface.

3. The chordwise Mach number is the cosine of the sweep angle times the free-stream Mach number. For 25°, it is 0.9063, which multiplied by the Mach number of 0.8, would yield 0.725.

4. The shock wave angle becomes less as the Mach number goes higher. Actually, the Mach number of flow in a supersonic wind tunnel can be determined by measuring the shock wave angle.

5. Vortex generators impart more energy to the air so that the flow over the control surface is effective, which might otherwise be separated due to shock wave formation.

6. The principle of the area rule is that drag will be minimized in supersonic flight if the total cross-sectional area of an aircraft progresses from fore to aft in a smooth fashion, rather than having abrupt changes. It is applied by reducing fuselage cross-sectional area in the vicinity of the wing.

7. Mach 5 is an average speed of a range where flow characteristics change rather significantly from lower supersonic speeds.

8. A supersonic combustion ram jet is necessary to provide thrust at supersonic speeds, but it would have to be in some way combined with a turbojet for operation at lower speeds.

# Chapter 8

1. First determine the gross weight. The payload weight would be 170 lb/person, or 340 lb, plus the baggage weight of 75, which yields 415 pounds. Considering the empty weight ratio to be 0.55, the overall average, and the fuel weight ratio to be 0.15 for this category airplane, the gross weight would be the payload weight divided by 1 minus the empty weight ratio and the fuel weight ratio, or,

$$W_g = 415 / (1 - 0.55 - 0.15) = 1,383 \text{ lb}$$

Next, calculate the wing area and span. From the equation for average wing loading, W/S is equal to 2.24 times the cube root of the gross weight minus 6, or,

$$W/S = 2.24 (11.1 - 6) = 11.46 \text{ lb/sq ft}$$

The wing area, then, is the gross weight divided by the wing loading,

$$S = 1383 / 11.46 = 120 \text{ sq ft (rounded off)}$$

A typical aspect ratio for this kind of airplane would be 7. The wingspan is then the square root of aspect ratio times wing area, or the square root of 7 × 120, which is about 29 ft.

Now we can calculate the approximate power necessary. Because it is to be a fairly clean airplane (in order to get the cruise speed desired), and a fixed gear, we could guess at a parasite drag coefficient about the same as a Cessna 182 Skylane. From Table 3-1, this figure is 0.031, and, if it is a low wing, the span efficiency factor can be estimated as 0.6. The velocity in ft/sec is the speed in knots times the conversion factor of 1.69, or 219.7 ft/sec. We now have all the values to calculate power required at cruise altitude:

| | |
|---|---|
| Cd | = 0.31 |
| Density ( Fig.  A-5 ) | = 0.001927 slugs/cu ft |
| S | = 120 sq ft |
| V | = 219.7 ft/sec |
| W | = 1,383 lb |
| b | = 29 ft |

Inserting all these values into the equation given in the example for THP required, we get 86.2hp.

The power lost in the propeller requires more BHP than this; hence, assuming a propeller efficiency of 80 percent, we divide 86.2 by 0.80 to get the brake horsepower of 107.7. Because this is cruise power, we assume that we can get maximum cruise power of 75 percent at 7,000 ft, and 100 percent, or rated, power would be BHP at cruise divided by 0.75, or,

$$\text{BHPrated} = 107.7 / 0.75 = 143.6 \text{ hp}$$

Because we would not find an engine rated at exactly this power, we would probably opt for a 150-hp engine.

Finally, the range and fuel requirements can be determined. Because we estimated in the beginning that the fuel weight ratio was 0.15, the amount of fuel is 15 percent of our gross weight of 1,383 lb, or 207.5 lb (approximately 35 gal). Range is determined by multiplying fuel weight times cruise velocity and dividing by specific fuel consumption times brake horsepower, written in equation form as,

$$R \ (Wf \times V) \ / \ (SFC \times BHP)$$

Using 0.5 as a typical SFC, and the BHP for the cruise condition, this becomes,

$$R \ (207.5 \times 130) \ / \ (0.5 \times 107.7) = 500 \ nm$$

This figure is not quite the required 600 miles, but if we are willing to accept it in order to get the other performance with a moderate amount of power, we have our design. If not, one could increase the amount of fuel, and recalculate all the parameters, iterating until all the performance is met. Note that this is not the only correct answer. Many assumptions have to be made, and the outcome will depend on the exact choices.

2. The Arrow has a horizontal tail volume coefficient, as determined in the text, of 0.5. Using this coefficient for our airplane, the tail volume would be determined from the equation,

$$Tail \ Vol. = Vh \times S \times Cav,$$

where Cav is the average chord. From chapter 2, recall that aspect ratio (AR) is span divided by average chord. Thus, rearranging the equation to solve for the average chord gives,

$$Cav = b \ / \ AR = 29 \ / \ 7 = 4.14 \ ft$$

Therefore,

$$Tail \ Vol. = 0.5 \times 120 \times 4.14 = 248.4 \ cu \ ft$$

The horizontal tail area is the tail volume divided by the moment arm, or

$$Sh = 248.4 \ / \ 12 = 20.7 \ sq \ ft$$

3. Because vertical tail volume ranges from approximately 0.02 to 0.05, an average value would be 0.035. Vertical tail volume is given by the equation,

$$Tail \ Vol. = Vv \times S \times b = 0.035 \times 120 \times 29 = 121.8 \ sq \ ft$$

Again using the same tail moment arm, the vertical tail area is the tail volume divided by this arm, or,

$$Sv = 121.8 \ / \ 12 = 10.2 \ sq \ ft$$

We now have our airplane pretty well sized from a very preliminary standpoint. The dimensions are:

Weight        = 1,383 lb
Wing area     = 120 sq ft
Wingspan      = 29 ft
Hor. tail area = 20.7 sq ft

Vert. tail area = 10.2 sq ft
Engine BHP   = 150
Fuel          = 35 gal

# Chapter 9

1. This arrangement minimizes the swirl from the propeller and the associated turbulence in the test section.

2. The wing area is 0.5 times 4, or 2 sq ft. From the appendix, the standard sea level density is seen to be 0.002377 slugs/cu ft. Dynamic pressure, q, is $1/2$ times the density times the velocity of 80 ft/sec squared, or 7.6 lb/sq ft. Dividing the lift by both q and wing area yields a lift coefficient of 0.5.

$$S = cb = 0.5 (4) = 2 \text{ sq ft}$$

$$q = 1/2 \text{ (density) } V^2 = 1/2 (0.002377) \, 80^2 = 7.6 \text{ lb/sq ft}$$

$$C_L = L/qS = 7.6/7.6(2) = 0.5$$

3. From the text, we see that Reynolds number at sea level is 6,329 times velocity times chord:

$$R = 6329 \, Vc = 6329(80)(0.5) = 253,160$$

This is a much smaller Reynold's number than normal full-scale flight conditions, which is more on the order of several million.

4. Force is usually measured with a balance, pressure is traditionally measured with a manometer, and flow visualization can be done with yarn tufts or smoke.

5. Shock waves can be visualized with a Schlieren apparatus, a device that blocks the deflected rays from a high intensity light source that pass through the denser area of the shock.

6. A simple procedure that would reveal the size of the wake from your fairing would be to tuft the fairing, and then observe it in flight. The amount of separated flow would determine the size of the wake, indicated by the tufts that flutter or point away from the surface. The smaller the wake that results, the smaller would be the pressure drag. If you cannot see the fairing from the cockpit, you may have to have a chase airplane do the observing.

# Bibliography

Abbott, I. H., Von Doenhoff, A. E., and Stivers, L. S., Jr., "Summary of Airfoil Data," NACA Report No. 824, Langley Aeronautical Laboratory, 1945.

Abbott, I. A., and Von Doenhoff, A. E., *Theory of Wing Sections*, McGraw-Hill, New York, 1949 (also Dover, New York, 1959).

Allen, Oliver E., *The Airline Builders*, Time-Life Books Inc., Alexandria, VA, 1981.

Anderson, John D., Jr., *Introduction to Flight*, McGraw-Hill, New York, 1978.

Anon., *Federal Air Regulations Part 23*, available from Superintendent of Documents, U.S. Government Printing Office, Washington, D.C., continuously updated.

Baals, Donald D. and Corliss, William R., *Wind Tunnels of NASA*, NASA, Washington, D.C., 1981.

Ball, Larry A., *Those Incomparable Bonanzas*, McCormick-Armstrong Co., Inc., Wichita, KS, 1971.

Bryan, C. D. B., *The National Air and Space Museum, Volume One Air*, Peacock/Bantam, New York, 1982.

Chant, Christopher, *Aviation, An Illustrated History*, Chartwell Books Inc., Orbis Publishing Limited, London, England, 1978.

Collins, Richard L., Editor, *Flying Buyers Guide 1983*, Ziff-Davis, New York, 1983.

Corning, Gerald, *Supersonic and Subsonic Airplane Design*, published by author, Box 14, College Park, MD, 1960.

Crawford, Donald R., *A Practical Guide to Airplane Performance and Design*, Crawford Aviation, P.O. Box 1262, Torrance, CA, 1981.

Dole, Charles E., *Flight Theory and Aerodynamics*, Wiley, New York, 1981.

Dommasch, D. O., Sherby, S. S., and Connolly, T. F., *Airplane Aerodynamics*, Pitman, New York, 1967.

Etkin, Bernard, *Dynamics of Flight-Stability and Control*, Wiley, New York, 1982.

Garrison, Paul, *The Complete Guide to Single-Engine Mooneys*, TAB Books, Inc., Blue Ridge Summit, PA.

Hallion, Richard P., *Designers and Test Pilots*, Time-Life Books, Inc., Alexandria, VA, 1983.

Hoerner, S. F., *Fluid-Dynamic Drag*, published by author, Midland Park, NJ, 1965.

Jones, R.T., *Wing Theory*, Princeton University Press, Princeton, NJ, 1990.

Kohn, Leo J., *The Flying Wings of Northrop*, Aviation Publications, Appleton, WI, 1977.

McCormick, Barnes W., *Aerodynamics, Aeronautics and Flight Mechanics*, Wiley, New York, 1979.

Misenhimer, Ted G., *Aeroscience*, Aero Products Research, Inc., Los Angeles, CA, 1973.

Nicolai, Leland M., *Fundamentals of Aircraft Design*, METS, Inc., 6520 Kingsland Court, San Jose, CA, 1975.

Oakes, Claudia M., *Aircraft of the National Air and Space Museum*, Smithsonian Institution Press, Washington, D.C., 1981.

Pazmany, L., *Light Airplane Design*, published by author, P.O. Box 10051, San Diego, CA, 1963.

Perkins, C. D. and Hage, R. E., *Airplane Performance, Stability, and Control*, Wiley, New York, 1949.

Pope, Alan, *Wind Tunnel Testing*, Wiley, New York, 1954.

Raymer, Daniel P., *Aircraft Design: A Conceptional Approach*, AIAA Education Series, AIAA, Washington, DC, 1989.

Reynolds, P. T., "The Learjet Longhorn Series—The First Jets with Winglets," Paper No. 790581, SAE, Warrendale, PA, 1979.

Smith, Hubert, *Performance Flight Testing*, TAB Books Inc., Blue Ridge Summit, PA, 1982.

Talay, T. A., *Introduction to the Aerodynamics of Flight*, NASA, Washington, D.C., 1975.

Taylor, John W. R., *Janes' All The World's Aircraft*, Franklin Watts Inc., New York, various editions.

Taylor, John W. R., *The Lore of Flight*, Crescent Books, New York, 1976.

Thurston, David B., *Design for Flying*, McGraw-Hill, New York, 1978.

Torenbeek, Egbert, *Synthesis of Subsonic Airplane design*, Kluwer Boston, Inc., Hingham, MA, 1982.

Wood, K. D., *Aerospace Vehicle Design Volume I Aircraft Design*, Johnson Publishing Co., Boulder, CO, 1963.

# Index

## A

absolute ceiling, climb performance, 124

accelerated climb performance, 152-156
  climb path optimization, 155-156
  energy height, 153-155
  rate of climb, 152
  specific excess power, 152

acceleration, 85
  accelerated climb performance, 152-156
  deceleration and landing performance, 144
  g-forces, 149
  takeoff performance, 139-140

adverse yaw, 176-177

aerodynamic center of airfoils, 27-28

aerodynamic principles vs. aerostatic principles, 10

aerodynamic testing, 247-270
  flight testing, 247, 266-270
  flow pattern testing, 259-262, 268, 269
  force testing, 255-257
  laboratory testing, 247
  pressure testing, 257-259
  problems in wind-tunnel testing, 264
  scale effect, wind-tunnel testing, 265-266
  Schlieren photography, 192, 264-265
  shock wave visualization testing, 263
  tufting for flow pattern testing, 259-262, 268, 269
  wall effects, wind-tunnel testing, 265
  wind tunnels (*see* wind tunnels)

aerodynamic twist, 45

aerospace plane (*see* hypersonic flight)

Aerospatiale TBM700, *295*

aerostatics, 3-7
  aerodynamics vs., 10
  air pressure, 6-7
  buoyancy principle, 3-6
  density of air, 6-7

ailerons, 42-43, 169-171

air pressure, 6-7
  Bernoulli principle, 13-20, 48
  center of pressure of airfoils, 26-27
  compressible flow, 188
  differential air pressure, 25, 36, 37
  dynamic pressure, 13-15
  friction effects, 28
  incompressibility of air, 188
  mass flow, acceleration, 86
  negative air pressure, 25, 36, 37
  pitching moment, 27-28
  positive air pressure, 25, 36, 37
  pressure distribution over airfoils, 24-28
  pressure drag, 56, 61
  pressure gradients, 28
  pressure gradients, adverse pressure gradients, 62
  static pressure, 13-15

airfoils (*see also* propellers; wings), 223-225, 289-291
  1-series airfoils, 35
  aerodynamic center, 27-28
  aerodynamic twist, 45
  ailerons, 42-43
  angle of attack, 21, 22-23, 31-33
  aspect ratio of wings, 39-40
  Bernoulli principle, 19-20, 48
  camber, 21
  cambered airfoils, lift, 22-23
  center of pressure, 26-27

chord, 21

chordline, 21

Clark Y, 33

design point, 223

development and design of airfoils, 31-36, 289-291

differential air pressure, 25, 36, 37

downwash effects, 38-39, 80-83

Eppler method, 289-290

Eppler, Richard, 289

flaps effect, 48-52

friction, 28

GA(W) series airfoils, 36, 200

glider design, 31-33

Gottingen G-398, 33

gradients, pressure gradients, 28

high-speed flight, 199-200

inverse design, 289

inverted airfoils, lift, 23-24

laminar flow, laminar boundary layer, 58-60, 200, 290

leading edge, 21, 31-33

lift and momentum-change considerations, 47-48

lift coefficient, 35, 45-47, 50, 51, 223, 257

lift created by airfoils, 17-20, 31

lift/drag (L/D) ratio, 223-225

maximum camber, 21

meanline or midline, 21

moment or torque, 27

NACA 6-series, 35, 223

NACA designs, five-digit, 34

NACA designs, four-digit, 33-34, 223, 301-309

negative air pressure effects, 25, 36, 37

negative lift, 24

NLF airfoils, 290, 291

pitching moment, 27-28

planform of wings, 36

bypass ratios, 92
climb performance, rate of climb, 146
decision speed, 148-149
endurance of aircraft, 146
equivalent shaft horsepower, 109
fuel consumption, 112-114, 226
hypersonic flight, 208-210
liftoff speed, 148-149
minimum control speed, 148-149
minimum unstick speed, 148-149
performance characteristics, 145-149
power-to-weight ratio, 225-226
pulsejets, 87-88
pure turbojets, 91, 111-112
ramjets, 87
range of aircraft, 146
SCRAMjets, 208-210
shaft horsepower, 109
specific fuel consumption (SFC), 112
stall speed, 148-149
takeoff performance, 148-149
thrust horsepower, 238-239
thrust required curve, thrust available curve, 145
thrust specific fuel consumption (TSFC), 112
turbofans, 91-93, 111
turbojets, 88-91, 109-110, 111
turboprops, 93, 109-110, 294
turboramjets, 210
turboSCRAMjets, 210
V-1 "Buzz Bomb," WW II, 88-89
venturi effects, 17-19
Joukowsky, flight theory, 11
Junkers, Hugo, cantilever wing design, 272

# K

kinematic viscosity, Reynolds numbers, 61
kitplanes, 297-298

# L

laboratory testing, 247
laminar flow, 58-60, 200, 290
Lanchester, Frederick, flight theory, 10-11
landing gear, 226-227, 272
landing performance, 142-145
    altitude and air pressure effects, 145
    brakes and braking forces, 144
    deceleration, 144
    distance to land, 143
    friction, coefficient of friction, 144
    lift/drag (L/D) ratio, 145
    net weight of aircraft, 144

runway conditions, 145
    stall speed, 143
    wind effects, 145
Langley, Samuel, 93-94, 158, 247
lateral axis, 159
lateral dynamic motion, 177-179
    Dutch roll, 179
    graveyard spiral, 179
    spiral divergence, 178-179
lateral stability, 159, 172-175
    dihedral effect, 174
    dihedrals, 173
    sideslip and fuselage effect, 174-175
    tail design, 173-174
Law of Conservation of Matter (see continuity equation, lift)
leading edge, airfoils, 21, 31-33
Lear Fan 2100, 295
Lear Jet, 197
Lear, William, 197, 274, 295-296
level flight performance, 115-119
lift, 12-55
    aerodynamic center of airfoils, 27-28
    aileron effect on wings, 42-43
    airfoils, 17-20, 31-36
    airfoils, cambered airfoils, 22-23
    angle of attack, 21, 22-23
    aspect ratio of wings, 39-40
    Bernoulli principle, 13-20, 48
    center of pressure of airfoils, 26-27
    coefficient of lift, 35, 45-47, 50, 51, 223, 257
    continuity equation, 15
    differential air pressure, 25, 36, 37
    downwash effects, wings, 38-39
    dynamic pressure, 13-15
    flaps effect, 48-52
    gradients, pressure gradients, 28
    gravity vs. lift, 12
    inverted airfoils, 23-24
    Law of Conservation of Matter, 15
    moment or torque, 27
    momentum-change considerations, 47-48
    negative air pressure, 25, 36, 37
    negative lift, 24
    pitching moment, 27-28
    positive air pressure, 25, 36, 37
    pressure distribution over airfoils, 24-28
    propellers, 101
    quantity of lift, 45-47, 50, 51
    slats and slots to control lift, 52-54
    spoilers to control lift, 54-55
    stagnation point, 22-23
    stalls, 28-31, 40-45
    static pressure, 13-15
    suction effect, 16
    symmetrical airfoils, 23-24, 31-33

takeoff performance, 140
    velocity, continuity equation, 15
    venturi effect, venturi tubes, 15-16
    washout in wings design, 43, 45
    weight vs. lift, 12
    wing design considerations, 223-225
    wing lift and span effects, 36-40
    zero lift, angle of zero lift, 47
lift coefficient, 35, 45-47, 50, 51, 223, 257
lift curve (see angle of attack; lift)
liftoff speed, jet engines, 148-149
Lilienthal, Otto, development, 8-9, 158
Lindbergh, Charles, 271, 273
liquid manometer, pressure testing device, 257-259
load factors
    performance, 149
    turning performance, 183-187
Lockheed Vega Winnie Mae, 273
longitudinal axis, 159
longitudinal stability, 159-162
    neutral stability, 161-162
    pitching moment vs. angle of attack, 161-162
    static longitudinal stability, 160
    trim condition, 160
LoPresti Piper SwiftFury, 298
LoPresti, Roy, design innovations, 297

# M

Mach numbers, 190-191
    critical drag Mach numbers, 196
    critical Mach numbers, 193
Mach waves, 192
Mach, Ernst, Mach numbers, 190-191
maintainability of design, 212
Manley, Charles, reciprocating engines development, 94
maneuvering performance, 149-152
    g-forces, 149
    load factors, 149
    maneuvering speed, 152
    pull-up maneuver, 149-150
    stall speed, 152
    V-g diagrams, 150
    V-n diagrams, 150-151
maneuvering speed, 152
mass flow, 86
Maxim, Howard, powered-flight development, 8
maximum camber, airfoils, 21
mean effective pressure, reciprocating engines, 96
meanline or midline, airfoils, 21
Messerschmitt Me-262, 90
minimum control speed, 148-149, 172

Whitcomb, Richard, 35-36, 200, 203-204, 286, 289
Whittle, Frank, turbojet development, 88
wind
  landing performance, 145
  range of aircraft, 134-135
  takeoff performance, 141
wind tunnels
  annular design tunnels, 251, 253
  closed-circuit designs, 250-251
  design and configurations, 249-254
  flow pattern testing, 259-262
  force testing, 255-257
  high-speed wind tunnels, 262-266
  history and development, 247-249
  open-circuit designs, 249-250
  pressure testing, 257-259
  problems with wind-tunnel testing, 264
  scale effect, 265-266
  Schlieren photography, 264-265
  shock wave visualization testing, 263
  spin tunnels, 253-254
  straight-through designs, 249-250
  supersonic tunnels, 262-263
  tests performed in wind tunnels, 254-262
  tufting for flow pattern testing, 259-262
  wall effects, 265
wing loading, 216, 238
wings (see also airfoils; propellers), 219-225, 237-238, 241
  aerodynamic twist, 45
  ailerons, 42-43, 169-171
  airfoil selection, 223-225

aspect ratios, 39-40, 238
basic configuration, 219-222
Bernoulli principle, 48
canards, 158, 277-280
cantilever designs, 221-222, 272
center of gravity effects, 165-167
design point, 223
differential air pressure, 36-37
downwash effects, 38-39, 80-83
drooped tip, 282-285
endplates, 285
filleting or fairing, 220
flaps effect, 48-52
flying wings, 280-281
high- vs. low-wing designs, 220-221
Hoerner wingtips, 282-285
lift, 36-45
lift and momentum-change considerations, 47-48
lift coefficient, 45-47, 50, 51, 223
lift/drag (L/D) ratio, 223-225
negative air pressure effects, 36-37
planform of wings, 36, 222-223
positive air pressure effects, 36-37
rectangular- vs. tapered-wing shape, 222-223
semispan, 40
slots and slats to control lift, 52-54
span loading, 292
spanwise lift and stall sequence, 40-45
spar locations, 225
spoilers to control lift, 54-55
stall speed calculations, 40-45, 238
struts and bracing, 221-222
swept-wing designs, 197-199, 288-289

thickness, 225
velocity, 47-48
vortex generators for high-speed flight, 201-202
vortices, 37-38
wake turbulence, 37-38
warping of wing, 48
washout, 43, 45
weight/wing area (W/S) ratio, 237-238, 241
wing loading, 216, 238
wing tip shape, effective aspect ratio, 282-285
winglets, 286-288
wingspan of wings, 36
wingtip vortices, 37-38
*Winnie Mae*, 273
Wood, K.D., 91
Wortmann, F.X., airfoil design, 289
Wright Brothers, 33, 271
  canard wing design, 158
  powered-flight development, 10-11
  propellers development, 101
  reciprocating engines development, 94
  stability in design, 158
  wind tunnel development, 248-249

# Y

yaw, 159
  adverse yaw, 176-177
  yaw controls, 170-171
  yaw instability, 178

# Z

zero lift, angle of, 47